Ordinary Differential Equations

Jae Kyong Cho

The great book of nature
is written in
the language of math.
Who am I?
I am Galileo Galilei.

CONTENTS

1. What is a Differential Equation? _1

2. Examples of Simple Differential Equations _10
Newton Mechanics _11
Newton's Law of Cooling _16
Libby's Radiocarbon Dating _22
Malthusian Law _30
Sublimation of a Spherical Substance _36
Dehydration of a Porous Substance _39

3. How to Solve First Order differential Equations _47

**4. Applications of the 1st Order Differential Equations II
   - Electric Circuits - _57**
Kirchhoff's Circuit Law _58
DC RL-Circuits _65
AC RL-Circuits _72
DC RC-Circuits _77
AC RC-Circuits _81
Discharge of a Capacitor _83

5. How to Solve 2nd Order Differential Equations _89
2nd Order Homogeneous Diff-eq. with Constant Coefficients _92
2nd Order Nonhomogeneous Diff-eq. with Constant Coefficients _109
2nd Order Diff-eq where Coefficients are not Constant _115
Laplace Transformations _126

**6. Applications of 2nd Order Differential Equations
   - Mechanical & Electrical Systems- _136**
Applications of 2nd Order Diff-eqs. with Constant Coefficients. _137
Applications of 2nd Order Nonhomogeneous Diff-eqs.
   with Const-Coeffs. I _162
Applications of 2nd Order Nonhomogeneous Diff-eqs.
   with Const-Coeffs. II _192
Applications of Systems of Diff-eqs_200

CHAPTER 1

Gosh! The midterm in engineering math is tomorrow, but the book never helps no matter how many times I read it.

Ah! I wish no engineering math in the world!
What am I to do if I sleep until the sun comes up?
ZZZ...

POP!

Hi, Ordy!

I am a guru of engineering math. I came to help you prepare for the exam.

Oh, such fabulous words to me!
Follow me!

少林寺
Oh, Dear!
Shaolin Temple!

172,
173

Hurry up! Chop-chop!
3 hundred times each!

BOOM...

BOOM...

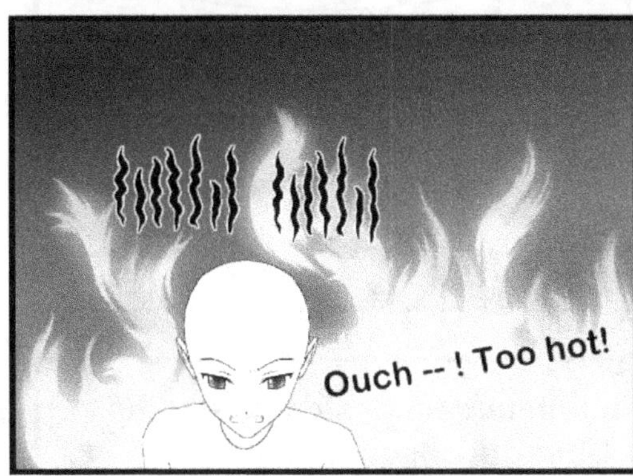
Ouch -- ! Too hot!

GYAHHH
Hot sand

See that?
See what?
?

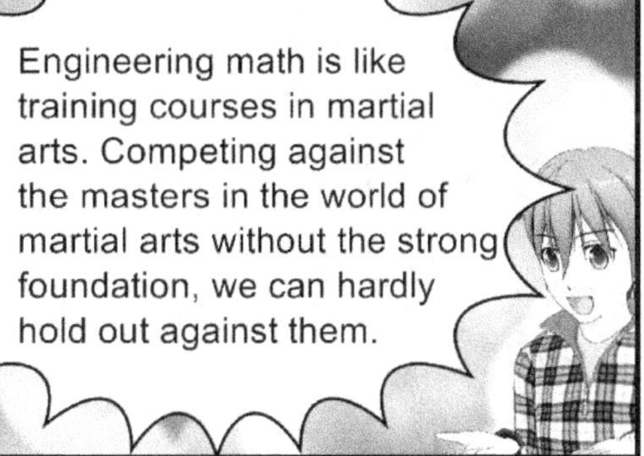
Engineering math is like
training courses in martial
arts. Competing against
the masters in the world of
martial arts without the strong
foundation, we can hardly
hold out against them.

In other words,
learning engineering math
can be compared to
learning the basics in martial arts.
The basics in martial arts
require fundamental physical
strength, skills in using hands
and feet, and etc.

Engineering Math =
Basics in Martial Arts

In engineering math, what comes
under the fundamental physical
strength, which is the most crucial,
then?

That's differential
equations!

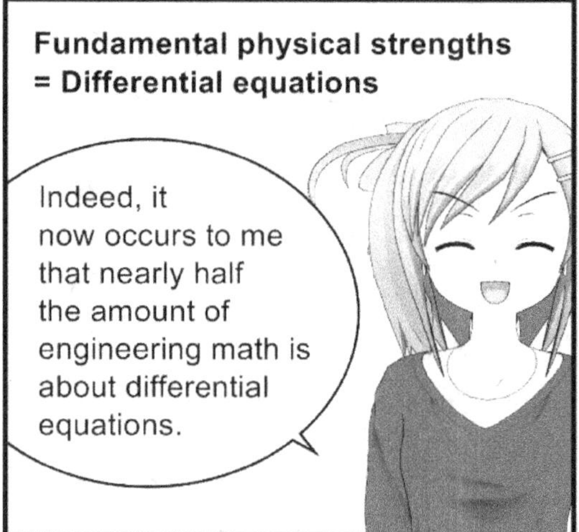
Fundamental physical strengths
= Differential equations
Indeed, it
now occurs to me
that nearly half
the amount of
engineering math is
about differential
equations.

The midterm coverage is differential equations too, isn't it!

My honorable guru!
Please, let me know about differential equations.
By all means, please!

Follow me around for just three days,

and you will be a guru of differential equations.

Why ever are differential equations so important?

That's because a state of any physical object in our world is determined by its velocity.

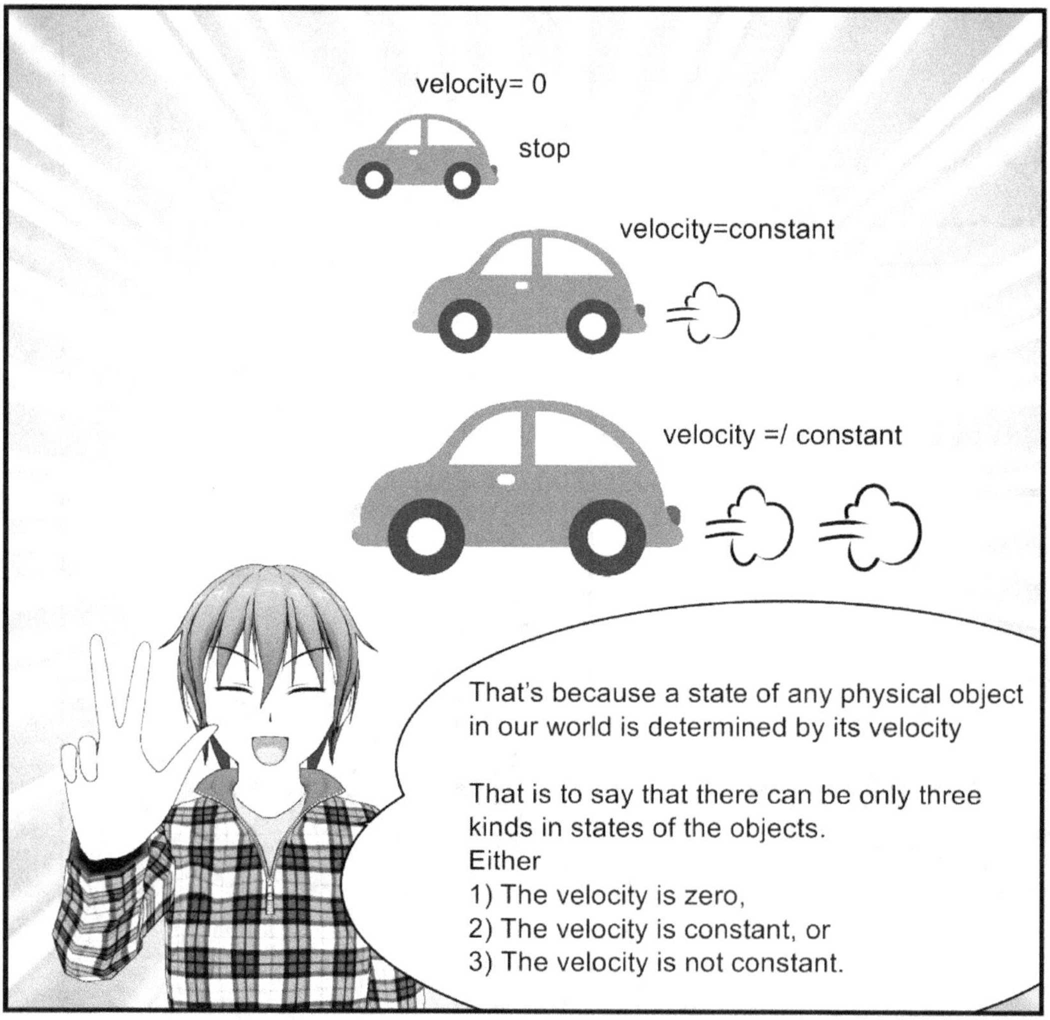

velocity= 0
stop
velocity=constant
velocity =/ constant
That's because a state of any physical object in our world is determined by its velocity

That is to say that there can be only three kinds in states of the objects.
Either
1) The velocity is zero,
2) The velocity is constant, or
3) The velocity is not constant.

Now, since a velocity (dL/dt) is a derivative of a distance (L) with respect to time (t), the expression includes a term of the derivative (dL/dt) and ends up with a differential equation.
Have you heard of 'Newton'?

Of course, I have!
I do know at least the Newton's first law, second law, and third law since I learned them when I was in high school!

Put aside the third law since it is about reciprocal actions between bodies, but let's consider the first and the second.
What is the first law?
1

The law says that if no force is added to a body, the body keeps maintaining its current velocity.
Smart am I, indeed!

That's correct, and it is nothing but the law of a constant velocity.
Of course, we can take the 0 velocity for a particular case of a constant velocity also.

What if we put the Newton's first law in an expression?
Since the velocity v is constant, we get v = dL/dt = k where k is a constant.
"dL/dt = k" is a differential equation!

What is the Newton's second low, then?
That is "F = ma."

Yes, it is the law of acceleration, which says a = dv/dt = F/m.
dv/dt = F/m, which is also a differential equation.
Newton

Hi, my dear brother!
You just have come at the right moment. Give her some explanations on the theory of relativity.
I am Einstein.

The theory of relativity explains the phenomenon that is obvious when an object is moving at a high velocity close to that of light.

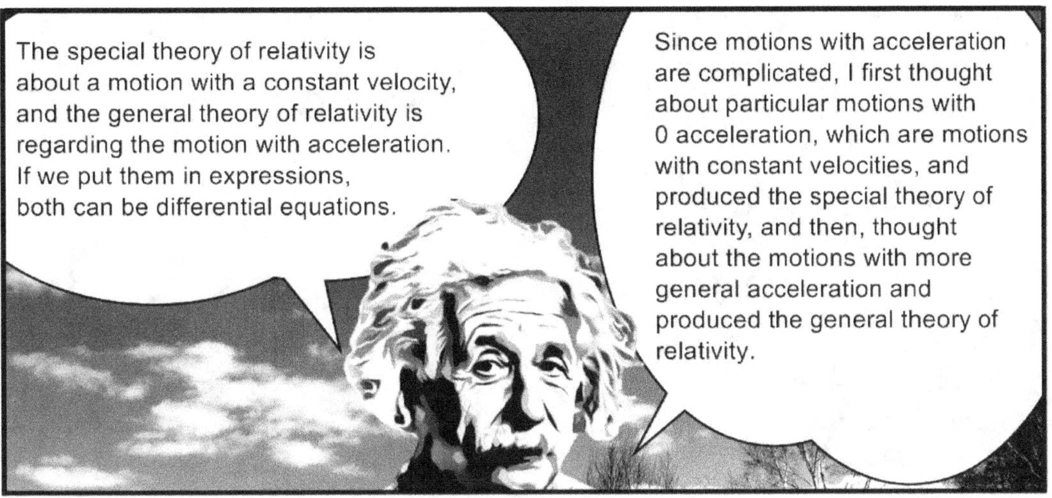
The special theory of relativity is about a motion with a constant velocity, and the general theory of relativity is regarding the motion with acceleration. If we put them in expressions, both can be differential equations.
Since motions with acceleration are complicated, I first thought about particular motions with 0 acceleration, which are motions with constant velocities, and produced the special theory of relativity, and then, thought about the motions with more general acceleration and produced the general theory of relativity.

Besides, the Maxwell's equation is a differential equation, and so is the Schrodinger's. Maxwell made a comprehensive compilation of theories in electromagnetism, and Schrodinger established quantum mechanics.
I can see that the critical expressions forming the skeleton of engineering are all differential equations.

Besides the expressions in engineering, if we put in expressions things in our interest, they all can be differential equations. That's because what we are interested in is change, the expressions include the derivatives, and eventually, differential equations result.

CHAPTER 2
Examples of
Simple Differential Equations

Newton Mechanics

The **Newton**'s second law says a = dv/dt = F/m. If we take the integrals of both sides of the equation since we need to find the velocity v,

we get $\int \dfrac{dv}{dt}dt = \int \dfrac{F}{m}dt$. Then, we get $v = \dfrac{F}{m}t + C$

where F = 5 N and m = 50 kg (my weight). Therefore, v(t) = (1/10)t+C.

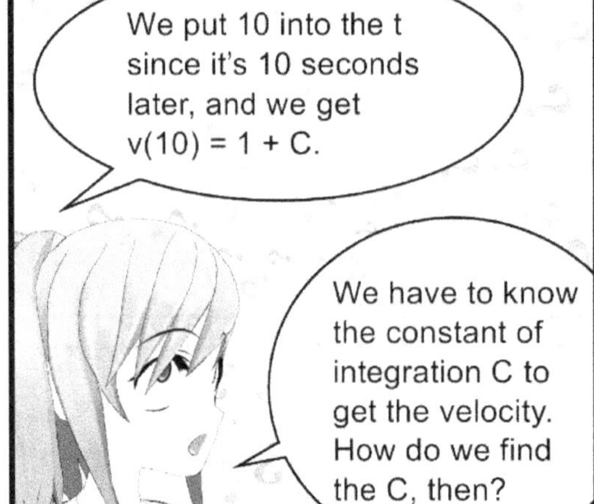

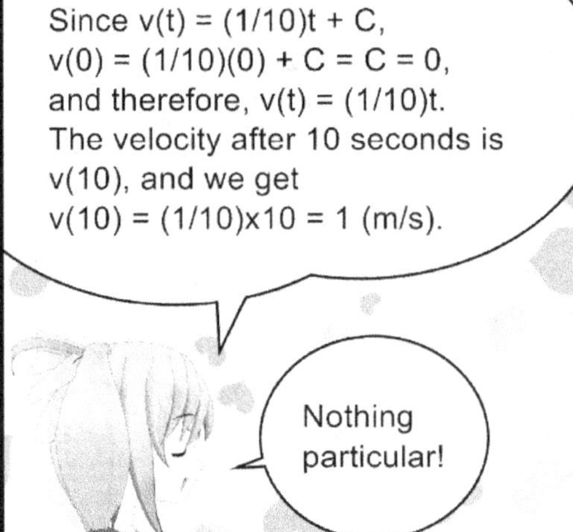

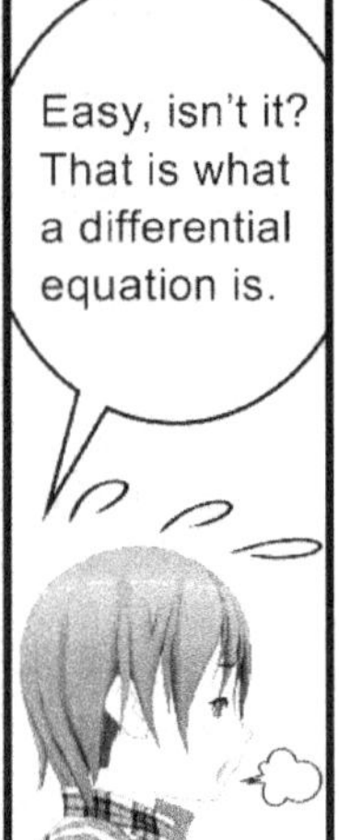

Where will you be located 10 seconds later, then?
10 seconds later
Get set! – Go!
L?

Just a piece of cake!
We get L = 10m
since the distance L covered is
v (1 m/s) X the time (10 seconds).
Forces keep getting added
and it's a motion
with acceleration,
you little beggar.
So, since the velocity changes,
we can't do the calculation
the way you just did!
Uh-uh, heck no!

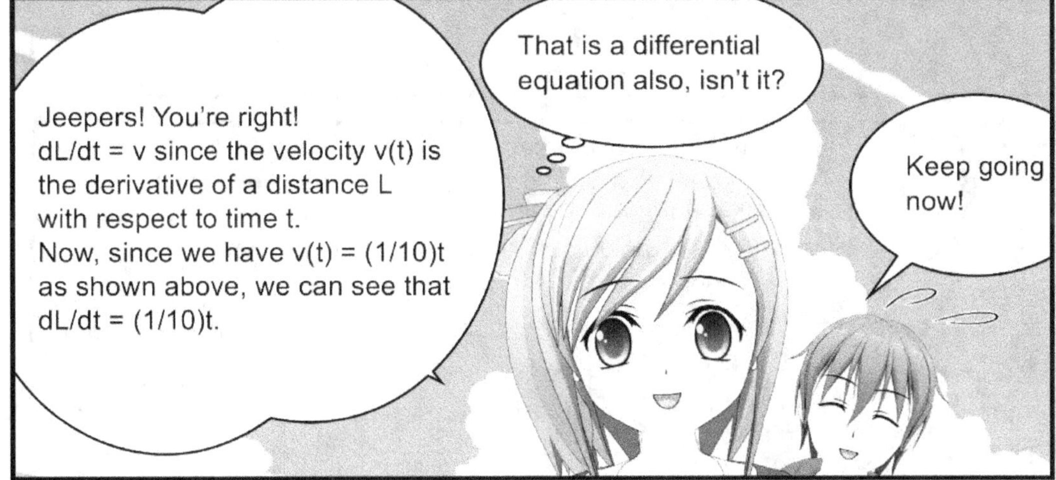

That is a differential
equation also, isn't it?
Jeepers! You're right!
dL/dt = v since the velocity v(t) is
the derivative of a distance L
with respect to time t.
Now, since we have v(t) = (1/10)t
as shown above, we can see that
dL/dt = (1/10)t.
Keep going now!

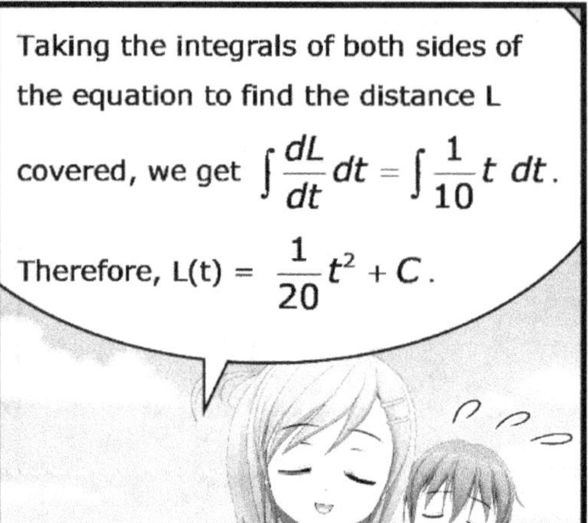
Taking the integrals of both sides of the equation to find the distance L covered, we get $\int \frac{dL}{dt}\,dt = \int \frac{1}{10}t\,dt$.
Therefore, $L(t) = \frac{1}{20}t^2 + C$.

Finding the constant of integration C by means of the initial condition $L(0) = 0$, we get
$L(0) = (1/20)(0)^2 + C = C = 0$.
Therefore, $L(t) = (1/20)t^2$.
Therefore after 10 seconds, I will be
$L(10) = (1/20)(0)^2 = 5$ (m) away.

Let's take a little break now!
Huff -- Huff!

Give me just one more push, please--!

By the way, didn't you get surprised?
No, but why?

You've still got a thickhead, indeed.

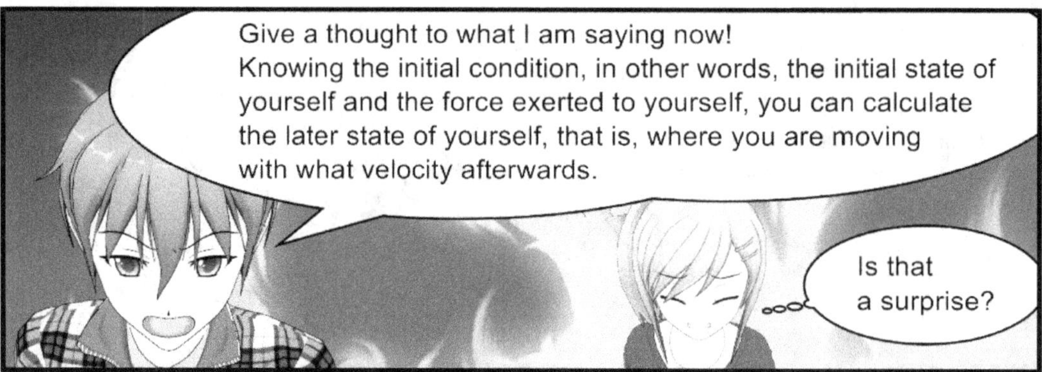
Give a thought to what I am saying now!
Knowing the initial condition, in other words, the initial state of yourself and the force exerted to yourself, you can calculate the later state of yourself, that is, where you are moving with what velocity afterwards.
Is that a surprise?

Have you ever thought about a destiny?

What destiny?
Do you mean such a statement that I will make my own fortune?

Fortunetellers believe that the moment of time of a human birth is an initial condition. Accordingly, they think knowing the initial condition, they can tell the fortune.
We call the idea fatalism or determinism, and may think it is based on the Newton Mechanics.
Among religions, some take such a position as above. Since human society has too many variables, fatalism doesn't work very well, though.
By means of differential equations, we can calculate the movements of all the planets in the universe including the earth's orbiting around the sun and motions of all the objects on the earth, can launch a rocket and have the spacecraft make a roundtrip to Mars, and etc.
Wow, marvelous! That's why Newton's famous!
My honorable guru! Let's go for a cup of coffee, now!

Newton's Law of Cooling

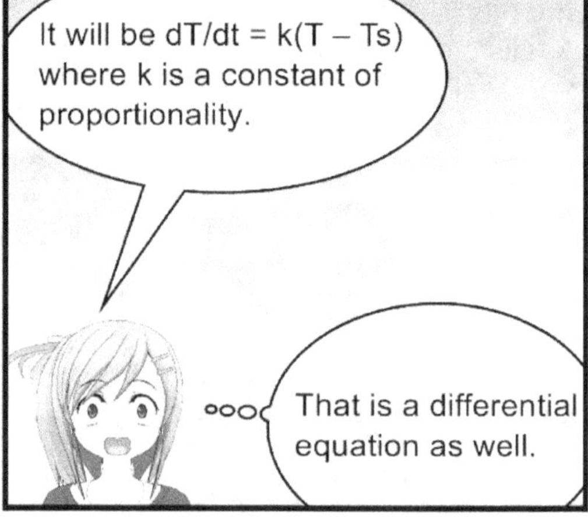

Assuming that coffee is in a room at 30°C and the coffee is now at 100°C, calculate the number of minutes required for the coffee to have the good temperature.

dT/dt = k(T − 30) since T_s = 30°C.
Taking the integrals of both sides of the equation to get the T(t), we get
$$\int \frac{dT}{dt}\,dt = \int k(T-30)\,dt \,.$$
T(t) = k(T − 30)t.
T(t) = -30kt/(1-kt).

Oh, my! You're talking nonsense!
The temperature T is a function of time t, which has to change along the flow of time, but if you take the integral of the right-hand side in such a manner that
$$\int k(T-30)\,dt = k(T-30)t \,,$$
then you've taken the T for a constant, haven't you?

How should I take the integral, then?

Take a look at the shelves over there. Got it?

Coffee cups and ice cream cups are assorted, aren't they?

That's all in a day's work.

Why don't you do the work in such a manner, then? You should assort the temperatures (T) and times (t), and then, take the integrals.

In dT/dt = k(T − 30), if we put together the temperatures on the left-hand side, we get dT/(T- 30), and then, if we move the time to the right-hand side, we get dT/(T − 30) = kdt. Now, we can take the integrals.

Right on! Keep going.
Seperation of variables

How do I find the antiderivatives from $\int \frac{dT}{T-30} = \int k\,dt$?

You did learn it when you were in high school, didn't you? If the deferential of the denominator is the numerator, the antiderivative is ln |the denominator|, and therefore, we get
$\int \frac{dT}{(T-30)} = \ln |T - 30|.$

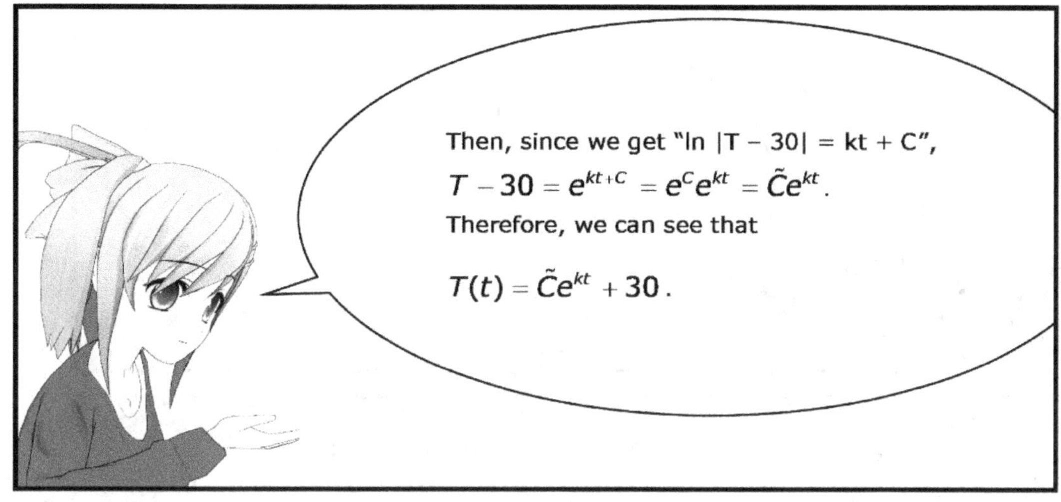

Then, since we get "ln |T − 30| = kt + C",
$T - 30 = e^{kt+C} = e^C e^{kt} = \tilde{C}e^{kt}$.
Therefore, we can see that
$T(t) = \tilde{C}e^{kt} + 30$.

Next, we need to find the constant $\tilde{C}$ ...
I got it! I can use the initial condition for that.
Initially (t = 0), the coffee is at 100°C,
that is, T(0) = 100, and therefore, we get
$T(0) = \tilde{C}e^{k0} + 30 = 100 \rightarrow \tilde{C} = 70$.
Therefore, we get
$T(0) = 70e^{kt} + 30$.

Since we need to find the time required
for the coffee to cool down to 70°C,
we set T(t) = 70, and then, find the time t.
$T(10) = 70 = 70e^{kt} + 30$
HA HA HA

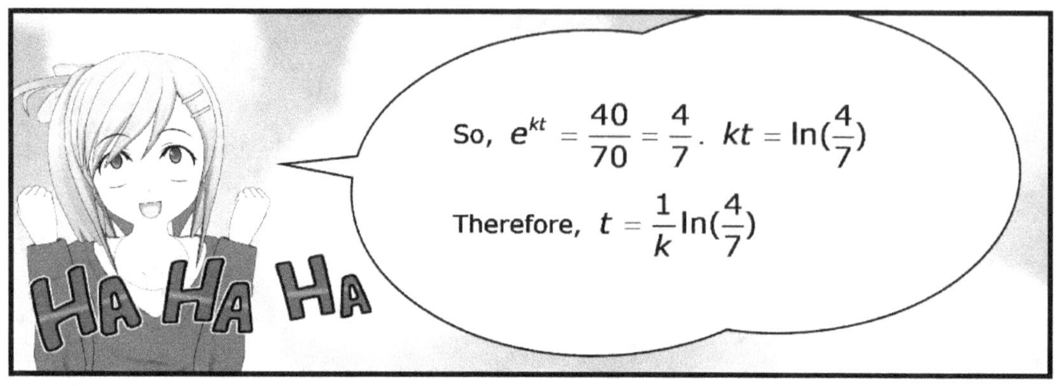

So, $e^{kt} = \dfrac{40}{70} = \dfrac{4}{7}$. $kt = \ln\left(\dfrac{4}{7}\right)$
Therefore, $t = \dfrac{1}{k}\ln\left(\dfrac{4}{7}\right)$

Oops! I have to know the k, don't I? How do I find the k, then?

My honorable guru! How do we find the k?

Give it a careful thought!
H'm—

I can see that we can never find it. The problem should have given another condition.

Shoot! Stop fooling me!

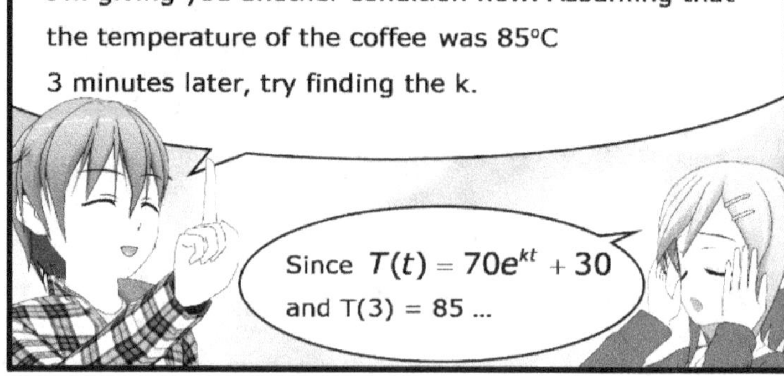

I'm giving you another condition now. Assuming that the temperature of the coffee was 85°C 3 minutes later, try finding the k.
Since $T(t) = 70e^{kt} + 30$ and $T(3) = 85$ ...

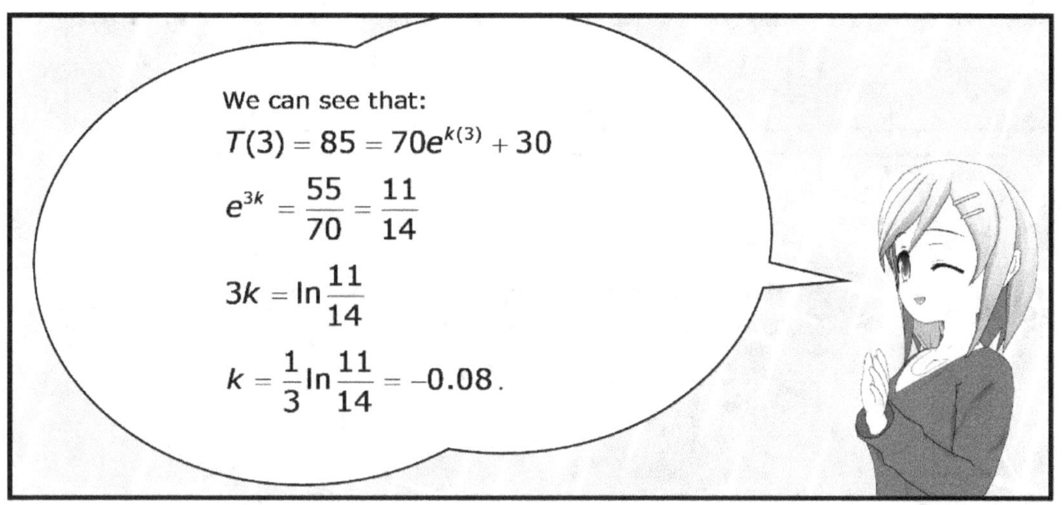

We can see that:
$T(3) = 85 = 70e^{k(3)} + 30$
$e^{3k} = \dfrac{55}{70} = \dfrac{11}{14}$
$3k = \ln\dfrac{11}{14}$
$k = \dfrac{1}{3}\ln\dfrac{11}{14} = -0.08$.

There we go! Keep going!

Consequently, $T(t) = 70e^{0.08t} + 30$. Since we are to find the time required for the temperature to fall to 70°C, we need to set $T(t) = 70 = 70e^{0.08t} + 30$, solve the equation for the t, and we can see that t ≈ 7 minutes.
YES!

You are good at arithmetic at the most. Now, let's go to beach for taking a break.

With the guru, not even one cup of coffee with style is ever possible.

Libby's Radiocarbon Dating
Decay of Radioactive Material

Oh!
A bone of
a dinosaur
OH!?

My honorable guru!
Here is a dinosaur's
bone.

How did you know
if it is a dinosaur's?

I just feel that it is.
Am I wrong?

Listen up, Ordy!
Science is not a field where we can make
a judgment based on our feelings.
We need to make an investigation first,
and then, should make a statement
with a sound basis.
On what do we
need to make
an investigation,
then?

To begin with, we need to
measure the ratio between
C^{14} and C^{12} inside this bone.

Why do we need
to take
the measurement?

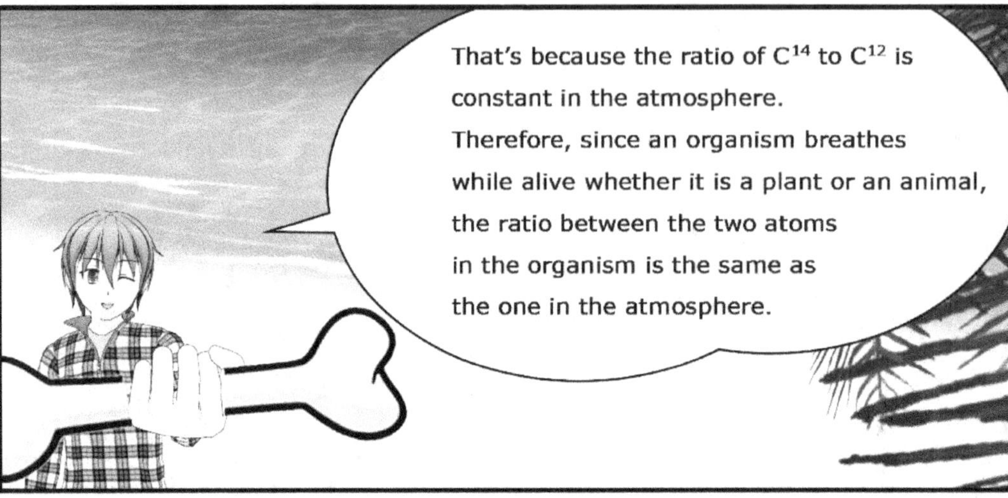
That's because the ratio of C^{14} to C^{12} is constant in the atmosphere.
Therefore, since an organism breathes while alive whether it is a plant or an animal, the ratio between the two atoms in the organism is the same as the one in the atmosphere.

Now, when it is dead, the carbon supplies by breathing or eating food get stopped.
Aha! Therefore, if it is dead, C^{14} decays since it is a radioisotope but C^{12} remains the same, and therefore, the ratio of C^{14} to C^{12} has to decrease as time passes away.

Therefore, if we know the decrease in percentage of the C^{14} inside the bone with respect to the C^{14} in the atmosphere, we can find the age of the bone.
Libby's won the Nobel Prize for the theory.
Wonderful!

Since a friend of mine Professor Joy has his office nearby, let's analyze the bone there.

Hi—
Professor Jones!

Wow, who is this!
Isn't that the guru?

What is the bone for?

We found it on the Beach.
Will you make some
analysis of it?

In this bone,
the remaining amount
of C^{14} is only 20%
of the original.

I've got to go for a meeting now,
but will be back soon.
Will you stay here for a little while
and wait for me, please?

OK! Have
the meeting
and come back.

Now, how do we calculate
the age of this bone?

We can calculate the time
required for the amount of
decrease of C^{14} to be 80%.
In what manner
does radioactive
material decay?

Um, the velocity of decay is
proportional to the current
amount of the mass.

What if we put the idea in a differential equation?

Taking y for the current amount of the mass, we get dy/dt = ky.

That's correct. Now, try solving that differential equation.

Since the cups at a coffee shop are assorted, we get dy/y = kdt.
Separation of variables!

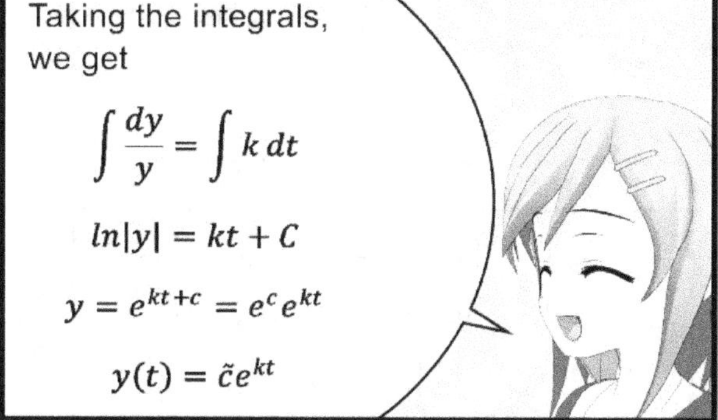
Taking the integrals, we get
$\int \frac{dy}{y} = \int k\,dt$
$ln|y| = kt + C$
$y = e^{kt+c} = e^c e^{kt}$
$y(t) = \tilde{c}e^{kt}$

Next, we need to find the constant C by means of the initial condition.

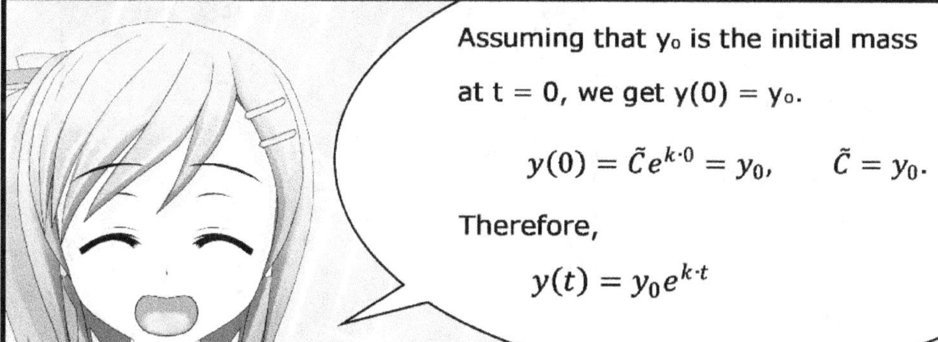
Assuming that y_0 is the initial mass at t = 0, we get y(0) = y_0.
$y(0) = \tilde{C}e^{k\cdot 0} = y_0, \qquad \tilde{C} = y_0.$
Therefore,
$y(t) = y_0 e^{k\cdot t}$

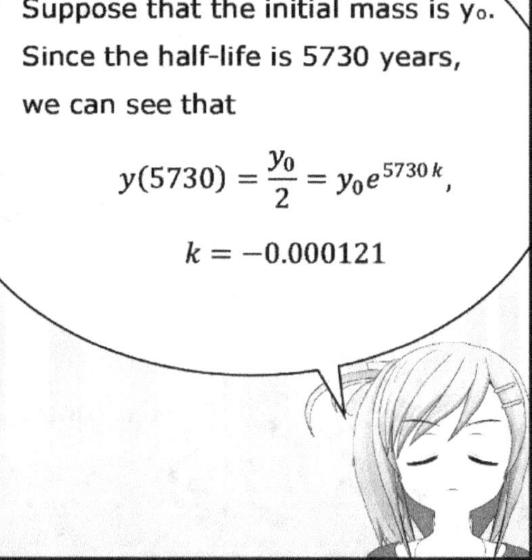

$$y(5730) = \frac{y_0}{2} = y_0 e^{5730\,k},$$

$$k = -0.000121$$

Therefore, we get
$y(t) = y_0 e^{-0.000121\,t}$.

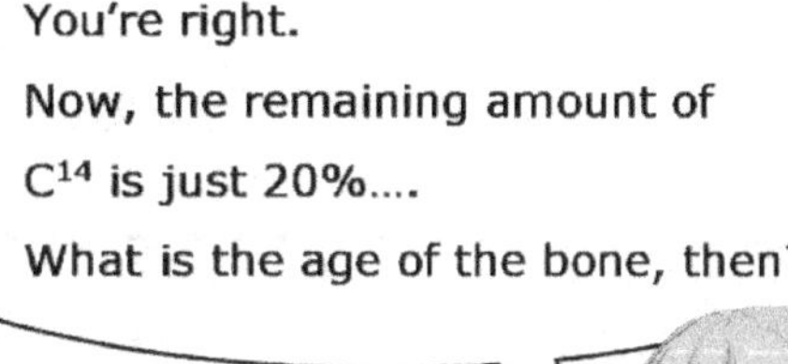

You're right.
Now, the remaining amount of C^{14} is just 20%....
What is the age of the bone, then?

Since the 20% means 1/5 of the original amount, we can set
$\frac{y_0}{5} = y_0 e^{-0.000121\,t}$
and solve the equation for the t, and then, we get t ≈ 13300 years.

Therefore, the age of the bone is approximately thirteen thousand and three hundred years.

Is this a bone of a dinosaur or not, then?

I am not sure ---

It's not a dinosaur's bone. That's because dinosaurs lived approximately from 200-million years ago to 60-million years ago.

Do we calculate the period of their existence by means of the content of the C^{14} in their fossils?

Of course!

Malthus's Law

…, 17, 18, 19, 20.
Quite a few more now!

What will be the total another two years later?

We can calculate that.

Can we calculate such a thing?

You betcha! Even our Ordy can do it. Have you heard of Malthus's Law?
Help --
No!

The Malthus's Law says that if a relatively small group is not disturbed significantly, the growth rate of the group is proportional to the current population of the group.
It will probably be a differential equation.

Hey Ordy, what if we put Malthus's Law in an expression?

Taking y for the current number of the goats, we get dy/dt = ky.
Separation of variables

What a smart girl!
Keep going now!

Solving the differential equation, we get
$\int \frac{dy}{y} = \int k \, dt$
$ln|y| = kt + C$
$y = e^{kt+c} = e^c e^{kt}$
$y(t) = \tilde{c} e^{kt}$
Huff Huff

Now that we have the solution to the equation, we need to find the constant $\tilde{C}$ by means of the initial condition.

Initially, we have t = 0 and y = 2 goats, and therefore, we get
$y(0) = \tilde{C} e^{k \cdot 0} = \tilde{C} = 2.$
Therefore,
$y(t) = 2 e^{kt}.$

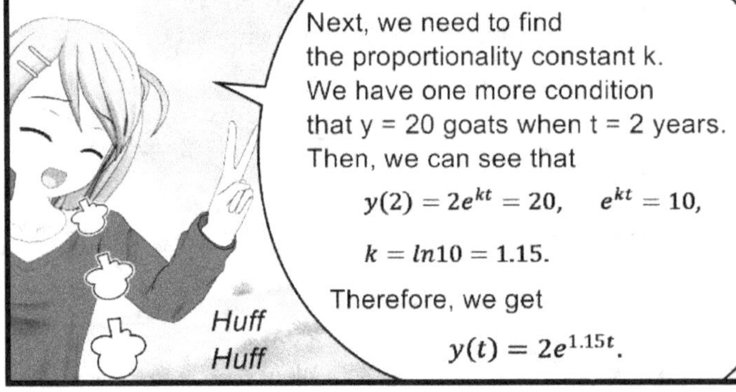
Next, we need to find the proportionality constant k. We have one more condition that y = 20 goats when t = 2 years. Then, we can see that
$y(2) = 2 e^{kt} = 20, \quad e^{kt} = 10,$
$k = ln 10 = 1.15.$
Therefore, we get
$y(t) = 2 e^{1.15t}.$
Huff Huff

When did the guru train a girl as that excellent student?

$$y(4) = 2e^{1.15 \times 4} = 190 \ goats$$

$$y(10) = 2e^{1.15 \times 10}$$
$$= 197{,}000 \ goats.$$

Bye! Come and visit again!
Bye now!

By the way, my honorable guru! I've found something very interesting.
What is it?

For both the decay of radioactive material and the number of the goats on the desert island, we get the same kind in differential equation.

Have you noticed that?
That's the fascination and greatness of math. Despite the fact that they are different phenomena from each other, they can be the same if we put them in mathematical terms.
The fascination of math! It's magical!

Another example is that the expressions for gravity, electric force, and magnetic force are the same to be

$F = k\dfrac{m_1 m_2}{r^2}.$

m_1
m_2
r
m_1
m_2
r
+
−
m_1
m_2
r
S
N
S
N

Hey Ordy!
Want to come and
visit my house?

I'd love to!

That's my house.

Sublimation of a Spherical Substance

Hey Ordy! How many months will it take for this mothball practically to disappear, say until the diameter falls to 1mm?

As expected, indeed!

Suppose that the velocity at which the volume of the mothball decreases due to its evaporation is proportional to the current amount of the surface of the ball.

Since dV/dt the velocity at which the volume decreases is proportional to the surface area S, we get dV/dt = kS.

Now, assuming that D is the diameter, we get

$$V = \frac{4}{3}\pi\left(\frac{D}{2}\right)^3, \qquad S = 4\pi\left(\frac{D}{2}\right)^2.$$

Since we need to set up a differential equation for the diameter, if we put them into the equation above, we get

$$\frac{d\left(\frac{4}{3}\pi\left(\frac{D}{2}\right)^3\right)}{dt} = k\left(4\pi\left(\frac{D}{2}\right)^2\right), \qquad \frac{4}{3}\pi\frac{3}{8}D^2\frac{dD}{dt} = k\left(\frac{4\pi}{4}D^2\right).$$

Therefore, we get

$$\frac{dD}{dt} = 2k.$$

Now, she is doing it all herself. My teaching is not in vain at all.

Solving the differential equation, we get

$\int dD = \int 2k\,dt$. D = 2kt + C.

Then, using the initial condition, we need to find the constant C.
Since the diameter of the mothball began with 2cm, we can set
D(0) = 2k•0 + C = C = 2.
Therefore, we get
D(t) = 2kt + 2.

Then, we need to find the constant k.
We've got a condition that D = 1cm at t = 3 months, and we can set
D(3) = 2k•3 + 2 = 1.
So, we get k = -1/6.
Therefore,
D(t) = 2(-1/6)t + 2 = (-1/3)t + 2.

Lastly, we need to find the time required for the diameter to be 1mm, and set
D(t) = 0.1 = (-1/3)t + 2.
Therefore, t = 5.7 months, isn't it?

There we go! I can see
your brain start working now.
It's late now, and that's it for today.

Dehydration of a Porous Substance

Wow, such a great dish! Ordy's cooking is nothing ordinary at all!
Burp—

HA HA HA
I am good at everything but engineering math.

Whoop! The sun is blazing now! It's gorgeous today!
The sunshine is so great that the comforter is already dried half in just an hour.

How long does it take for the comforter to be dried enough?

Let me see!
Golly! All the math has to begin this morning, doesn't it? Is this domestic math?

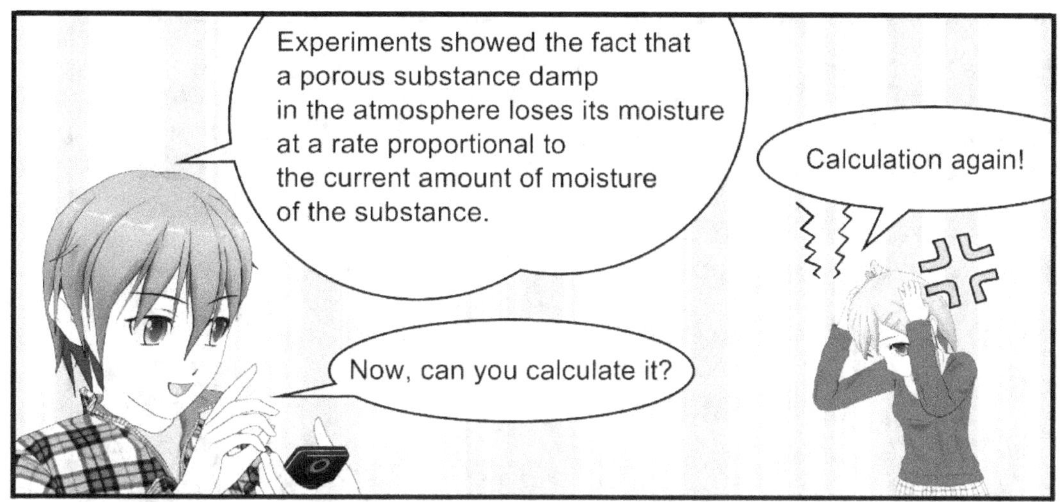

Experiments showed the fact that a porous substance damp in the atmosphere loses its moisture at a rate proportional to the current amount of moisture of the substance.
Calculation again!
Now, can you calculate it?

Um, assuming that the current moisture content of the comforter is y, we get
$\frac{dy}{dt} = ky$
since the rate of change in the moisture content $\frac{dy}{dt}$ is proportional to the y.
Uh-oh! I've got the same expression again!

Proceed, now!
Separation of variables
Solving the differential equation, we can see that
$\int \frac{dy}{dy} = \int k\,dt$
$\ln |y| = kt + C$
$y = e^{kt-C} = e^C e^{kt} = \grave{C} e^{kt}$

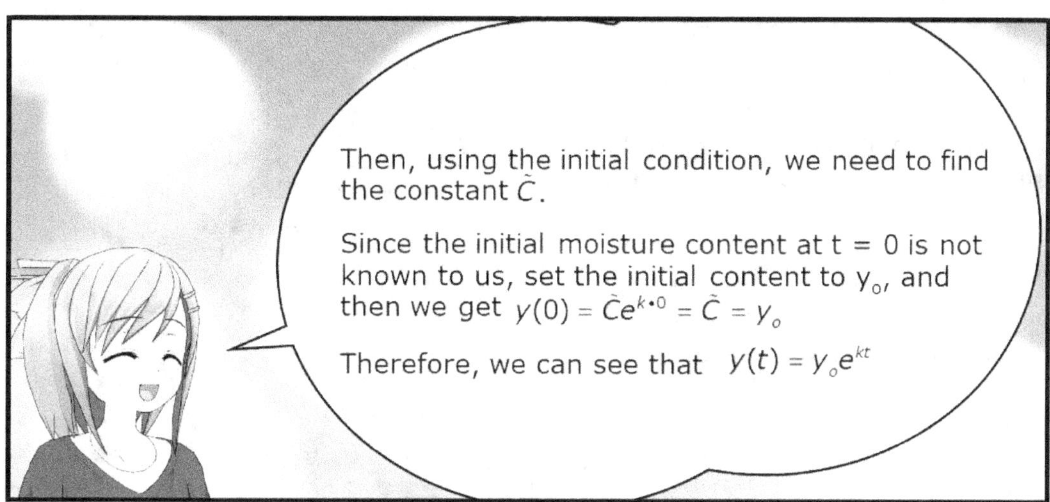

Then, using the initial condition, we need to find the constant $\grave{C}$.
Since the initial moisture content at t = 0 is not known to us, set the initial content to y_o, and then we get $y(0) = \grave{C} e^{k \cdot 0} = \grave{C} = y_o$
Therefore, we can see that $y(t) = y_o e^{kt}$

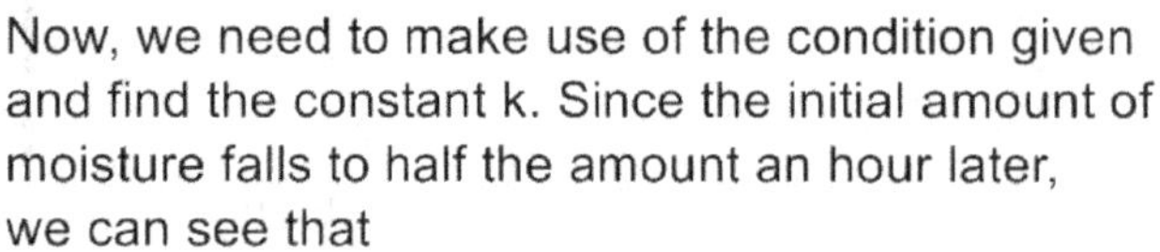

$$y(1) = y_o e^{k \cdot 1} = 0.5 y_o \, ,$$

$$e^k = 0.5 \, ,$$

$$y(t) = y_o e^{-0.69t} \, .$$

$$0.01 y_o = y_o e^{-0.69t} \, ,$$

The fun of science is in finding not facts
but the new ways
in which we can think about the facts.
I am W.L. Brag,
worked with my father
on structures in crystals
using X-rays,
and won a Nobel Prize
in Physics.

Quiz 1

Suppose that we put water into a water tank so that the depth of water is h. When we open the tap as shown in the figure, the velocity at which the depth decreases as time proceeds is proportional to the square root of the depth h. Put the relationship between the depth and the velocity in a differential equation.

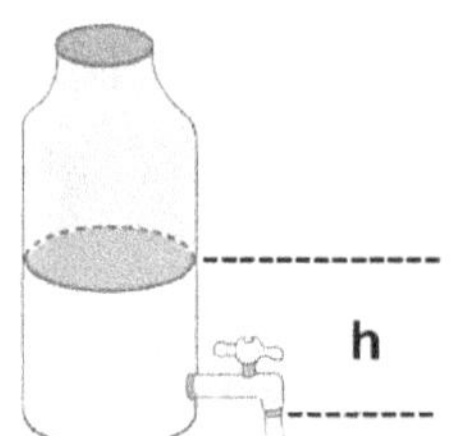

Answer

$$\frac{dh}{dt} = -k\sqrt{h} \text{ , (k is a proportionality constant)}$$

Quiz 2

When a skydiver with a parachute open falls at a velocity of v, the applicable forces are the gravity **mg** and the air resistance −bv² (b is constant). Set up a differential equation for the velocity v.

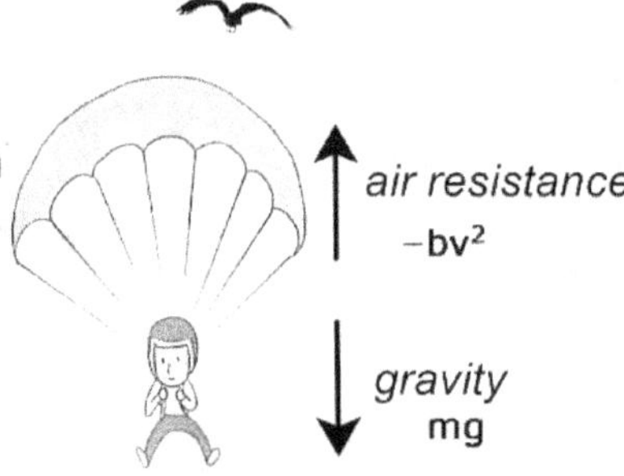

Answer

Newton's 2nd law says

$$F = ma = m\frac{dv}{dt} \text{ where F is net force, F = mg − bv}^2,$$

and therefore, $m\dfrac{dv}{dt} = mg - bv^2$.

Quiz 3

As a light from a lighthouse proceeds into a body of seawater, it gets absorbed partly by the water. The depth rate of change in the intensity of the light is proportional to the intensity at a given depth. Set up a differential equation for the intensity that is a function of the depth.

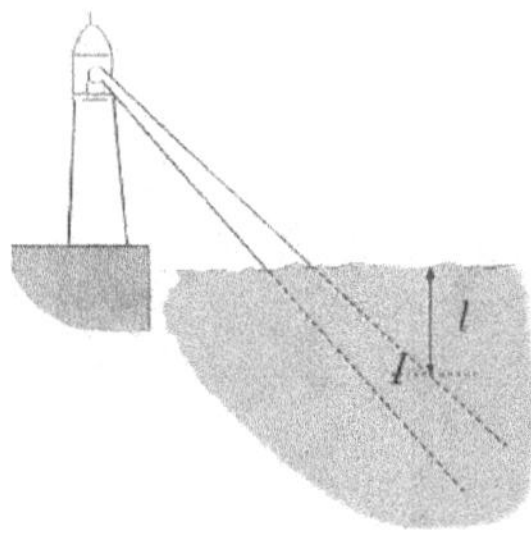

Answer

The depth (L) rate of change in the intensity I is $\dfrac{dI}{dL}$, which is proportional to the I.

Therefore, $\dfrac{dI}{dL} = kI$ (k is a proportionality constant).

Quiz 4

Suppose that at time t, both the birth rate and the death rate
in a country are proportional to the population at the time t.
(Note that the constants of proportionality are different.)
Assuming that there is neither immigrant nor emigrant,
set up a differential equation for the population p(t) at the time t.

Answer

The birth rate = ap (a: a constant), the death rate = bp (b: a constant),

and the time rate of change in population $\dfrac{dp}{dt}$

is 'the birth rate − the death rate'.

Therefore, we can put $\dfrac{dp}{dt} = ap - bp = (a - b)p$.

Quiz 5

Observation showed that the rate of change of
the atmospheric pressure P with altitude h is proportional to
the pressure at a given altitude.
Set up a differential equation for the pressuren P(h)
that is a function of the altitude.

Answer

The rate of change of the atmospheric pressure P
with altitude h is

dP/dh,

which is proportional to the pressure P.
Thererfore,

dP/dh = kP (k is a proportionality constant)

Quiz 6

Experiments showed that if a body falls in vacuum due to gravity, the acceleration is constant and is called gravitational acceleration, g.
a. Put the law in a differential equation for the velocity v(t).
b. Find the velocity of the body 3 seconds later.
c. Find the distance fallen of the body in 3 seconds

Answer

a. Since the acceleration is $\frac{dv}{dt}$, $\frac{dv}{dt} = g$.

b. By integration $\int \frac{dv}{dt} dt = \int g \, dt$, we get the velocity $v = gt + C$.

Since initial condition $v(0) = 0 = g(0) + C, C = 0, v(t) = gt$.

$v(3) = g(3) = (9.8 \text{ meters/sec}^2)(3 \text{ sec}) = 29.4 \text{ m/s}$

c. Since $v = \frac{dl}{dt} = gt$, $\int \frac{dl}{dt} dt = \int gt \, dt$, $l = \frac{1}{2}gt^2 + \tilde{C}$.

From initial condition $l(0) = 0 = \frac{1}{2}g(0)^2 + \tilde{C}, \tilde{C} = 0, l(t) = \frac{1}{2}gt^2$

$l(3) = \frac{1}{2}(9.8)(3)^2 = 44.1 \text{ m}$.

* In case of a free fall in vacuum, the acceleration is not subject to the mass of the falling object, and is constant($\because a = \frac{dv}{dt} = \frac{F}{m}$, where F is mg)

* This also applies to the free fall in air if we can neglect air resistance, for instance, if we drop an iron ball.

CHAPTER 3

Now, whatever the problem you may present, I'll solve it straight away.

They that know nothing fear nothing. You're kidding yourself!
HA HA HA

Still far away from the mastership!

What belt do I deserve in terms of Taekwondo?
What belt do you think other than the white?

Take it easy, though; it will be the yellow in no time.
Tell me any particular point common to the diff-eqs covered so far.
Well, they all have time derivatives?

Newton Mechanics: $\dfrac{dv}{dt} = \dfrac{F}{m}$

Newton's Law of Cooling: $\dfrac{dT}{dt} = k(T - T_s)$

Decay of Radioactive Substance: $\dfrac{dy}{dt} = ky$

Malthusian Law: $\dfrac{dy}{dt} = ky$

Mothball: $\dfrac{dV}{dt} = kS$

Comforter: $\dfrac{dy}{dt} = ky$

In the above, if r(x) = 0, that is,
if y' + p(x)y = 0
we call it the 1st order homogeneous diff-eq.
Homogeneous!

Solve the 1st order homogeneous diff-eq.

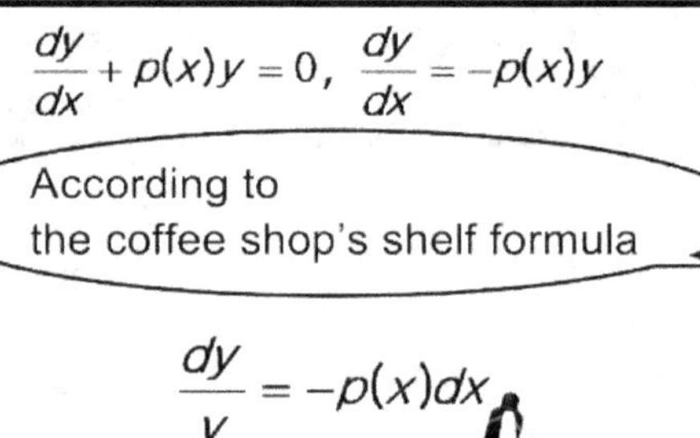

$\dfrac{dy}{dx} + p(x)y = 0,\quad \dfrac{dy}{dx} = -p(x)y$
According to
the coffee shop's shelf formula
$\dfrac{dy}{y} = -p(x)dx$

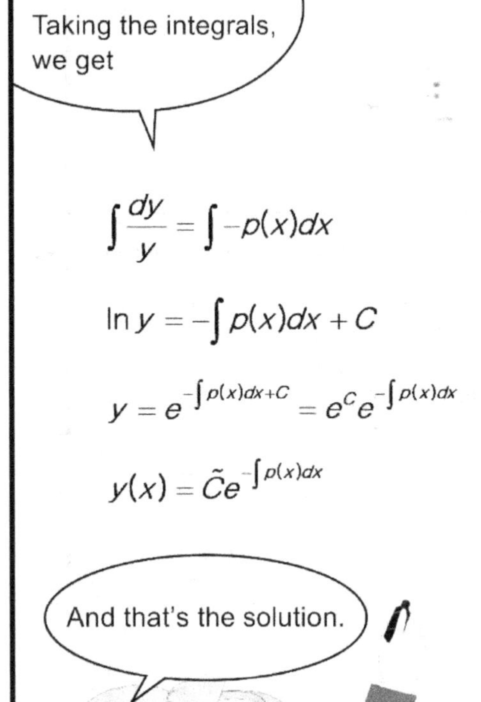

Taking the integrals, we get
$\displaystyle\int \dfrac{dy}{y} = \int -p(x)dx$
$\ln y = -\displaystyle\int p(x)dx + C$
$y = e^{-\int p(x)dx + C} = e^{C}e^{-\int p(x)dx}$
$y(x) = \tilde{C}e^{-\int p(x)dx}$
And that's the solution.

You just did a bang-up job!
Since $\tilde{C}$ is a constant also,
we just put $\tilde{C}$ = C.

And if we set
$\displaystyle\int p(x)dx = h$
we get
$y(x) = Ce^{-h}$

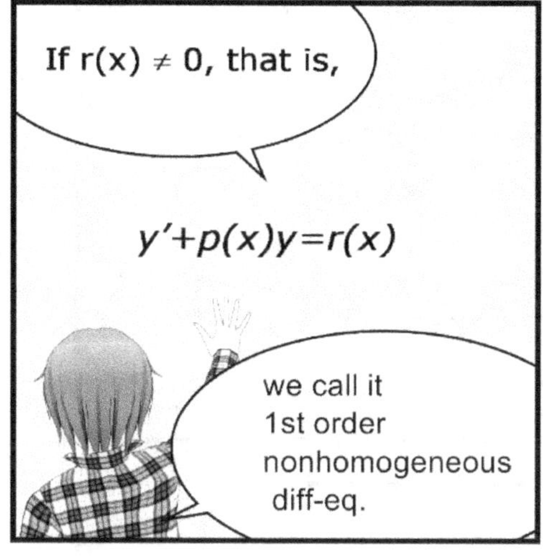

If r(x) ≠ 0, that is,
$y'+p(x)y=r(x)$
we call it
1st order
nonhomogeneous
diff-eq.

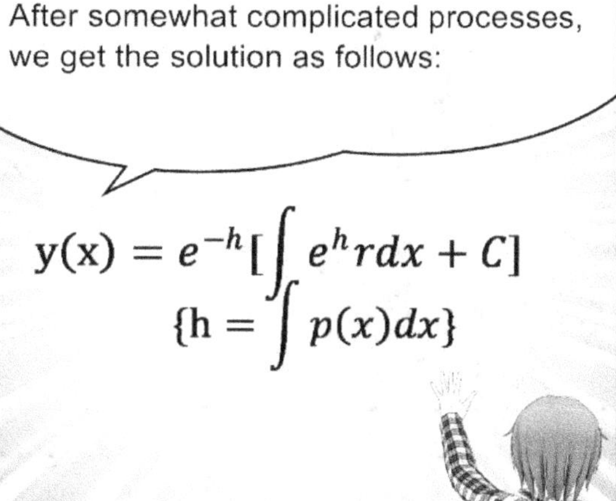

After somewhat complicated processes, we get the solution as follows:
$$y(x) = e^{-h}\left[\int e^{h} r\, dx + C\right]$$
$$\{h = \int p(x)\, dx\}$$

See that?
See what?

Since the solution is for the general form of 1st order diff-eq,
using the solution for a formula,
we can get a solution
to any diff-eq whether
homogeneous or not.

A ha!
It's similar to $x = \dfrac{-b \pm \sqrt{b^2 - 4ac}}{2a}$
the formula for the solution to the quadratic equation covered in high school.

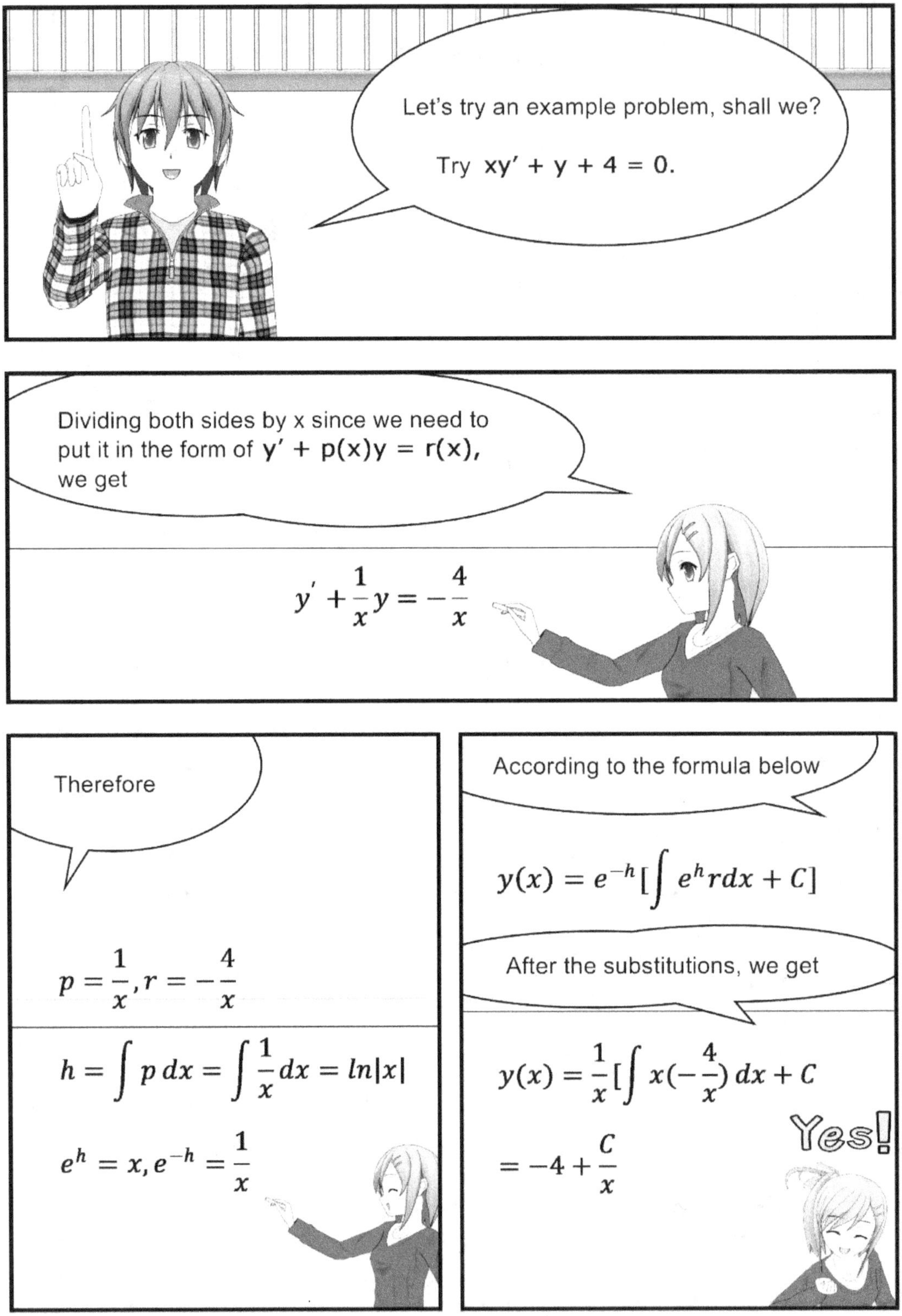

Dividing both sides by x since we need to put it in the form of $y' + p(x)y = r(x)$, we get

$$y' + \frac{1}{x}y = -\frac{4}{x}$$

Therefore

$$p = \frac{1}{x},\ r = -\frac{4}{x}$$

$$h = \int p\,dx = \int \frac{1}{x}\,dx = ln|x|$$

$$e^h = x,\ e^{-h} = \frac{1}{x}$$

According to the formula below

$$y(x) = e^{-h}\left[\int e^h r\,dx + C\right]$$

After the substitutions, we get

$$y(x) = \frac{1}{x}\left[\int x\left(-\frac{4}{x}\right)dx + C\right]$$

$$= -4 + \frac{C}{x}$$

Quiz 1

a. What is the general form of the 1st order diff-eq?

b. What is the formula for the general solution to the 1st order diff-eq?

Answer

a. $y' = p(x)y = r(x)$

b. $y(x) = e^{-h}\left[\int e^{h} \cdot r\, dx + C\right], \quad \left(h = \int p(x)\, dx\right)$

Quiz 2

Find the general solution to y' = ky.

Answer

Putting the equation in the general form of y' + py = r,
we get y' - ky = 0, and therefore, p = - k and r = 0.
Finding the h first to put it into the formula for the general solution,
we get

$$h = \int k\, dx = kx$$

Putting in the h, we get

$$y(x) = e^{-kx}\left[\int e^{kx} \cdot 0\, dx + C\right] = Ce^{-kx}$$

Quiz 3

Find the general solution to y' + 4y - 10 = 0.

Answer

Putting the equation in the general form of y' + py = r,
we get y' + 4y = 10, and therefore, p = 4 and r = 10.
Finding the h first to put it into the formula for the general solution,
we get

$$h = \int 4\, dx = 4x$$

Substituting h and r into the formula, we get

$$y(x) = e^{-4x}\left[\int e^{4x} \cdot 10\, dx + C\right] = e^{-4x}\left(\frac{10}{4}e^{4x} + C\right) = \frac{5}{2} + Ce^{-4x}$$

Quiz 4

Find the general solution to $2y' + 6y - e^{-3x} = 0$.

Answer

Putting the equation in the general form of y' + py = r, we get

$$y' + 3y = \frac{1}{2}e^{-3x}$$

and therefore

$$p = 3, r = \frac{1}{2}e^{-3x}$$

Finding the h first to put it into the formula for the general solution, we get

$$h = \int 3\,dx = 3x$$

Substituting h and r into the formula, we get

$$y(x) = e^{-3x}\left[\int e^{3x}\frac{1}{2}e^{-3x}\,dx + C\right] = e^{-3x}\left[\frac{1}{2}x + C\right]$$

Quiz 5

Find the general solution to y' + 4y - 8x = 0.

Answer

Putting the equation in the general form of y' + py = r, we get

$$y' + 4y = 8x$$

and therefore

$$p = 4, r = 8x, h = \int 4\,dx = 4x$$

Substituting h and r into the formula, we get

$$y(x) = e^{-4x}\left[\int e^{4x}8x\,dx + C\right]$$

$$= e^{-4x}\left[\int e^{4x}\,dx \cdot 8x - \int \{(\int e^{4x}\,dx \cdot (8x)'\}\,dx + C\right]$$

$$= e^{-4x}\left[\frac{1}{4}e^{4x} \cdot 8x - \int \frac{1}{4}e^{4x} \cdot 8\,dx + C\right]$$

$$= e^{-4x}\left[2xe^{4x} - \frac{1}{4}\cdot\frac{1}{4}e^{4x} \cdot 8 + C\right] = 2x - \frac{1}{2} + Ce^{-4x}$$

Quiz 6

Solve the initial value problem $2y' + 6y - e^{-3x} = 0, y(0) = 2.$

Answer

Putting the equation in the general form of y' + py = r, we get

$$y' + 3y = \frac{1}{2}e^{-3x}$$

and therefore

$$p = 3, r = \frac{1}{2}e^{-3x}, h = \int 3\,dx = 3x$$

Substituting h and r into the formula, we get

$$y(x) = e^{-3x}\left[\int e^{3x}\frac{1}{2}e^{-3x}dx + C\right] = e^{-3x}\left[\frac{1}{2}x + C\right]$$

Applying the initial condition

$$y(0) = 2 = e^{-3\cdot 0}\left[\frac{1}{2}\cdot 0 + C\right] = C$$

Therefore the particular solution satisfying the initial condition is

$$y(x) = e^{-3x}\left[\frac{1}{2}x + 2\right]$$

Quiz 7

Find the general solution to $y' = 8x^3y^2.$

Answer

Since it can't be put in the form of y' + py = r,

try the coffee shop formula (separation of variables).

Since y'=dy/dx

$$y' = \frac{dy}{dx} = 8x^3y^2.$$

By separating variables, we have

$$\frac{dy}{y^2} = 8x^3\,dx.$$

Taking the integrals of both sides, we get

$$\int \frac{1}{y^2}\,dy = \int 8x^3\,dx, \qquad -\frac{1}{y} = 2x^4 + C.$$

Therefore

$$y = \frac{-1}{2x^4 + C}.$$

Quiz 8

Find the general solution to yy′ = 1.

Answer

Since it can't be put in the form of y′ + py = r,

try the coffee shop formula (separation of variables).

Since y′=dy/dx ,

$$yy^{'} = y\frac{dy}{dx} = 1.$$

By separating variables, we have

$$ydy = dx.$$

Taking the integrals of both sides, we get

$$\int ydy = \int dx, \qquad \frac{1}{2}y^2 = x + C.$$

Therefore

$$y^2 = 2x + \tilde{C} \ .$$

CHAPTER 4
Applications of the 1st Order Diff-eq II
-Electric Circuits-

Kirchhoff's Circuit Laws

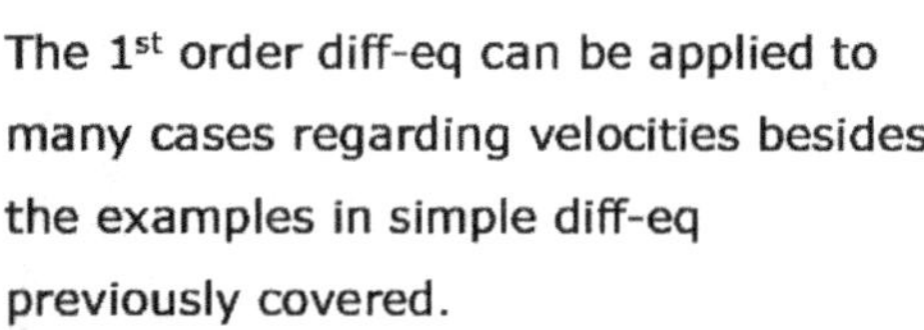

The 1st order diff-eq

$$y' + p(x)y = r(x)$$

Applications regarding velocities

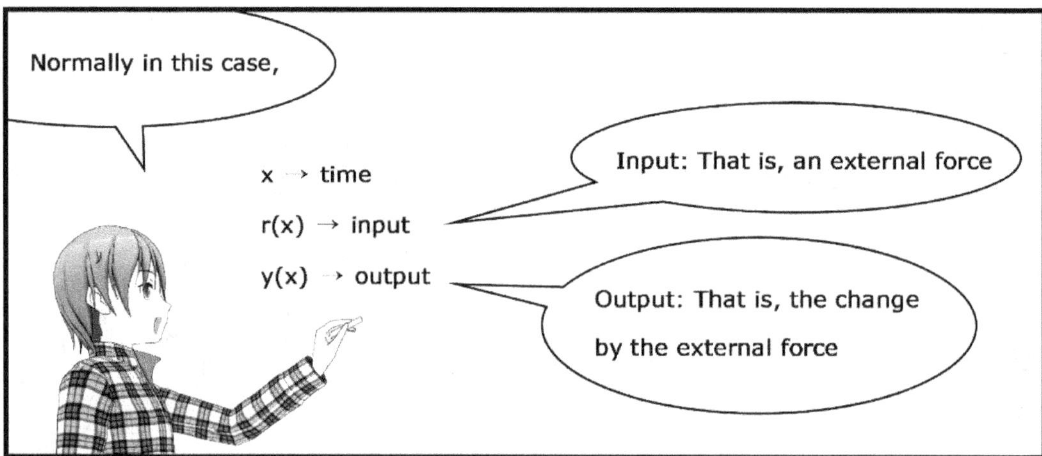

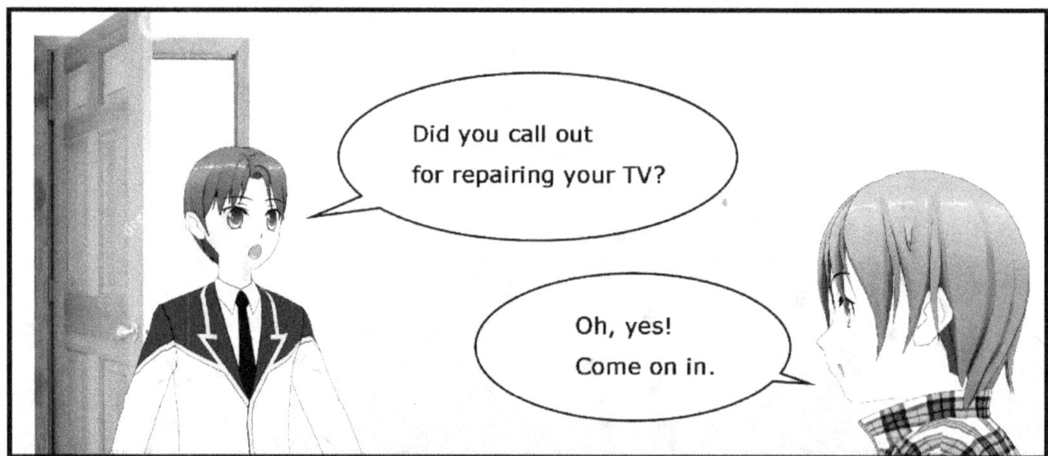

It's all done now.
Thanks. Bye!

Hey Emma, do you remember
the circuit diagram
the technician was looking at?

Yes

What was the first thing
you noticed?

took up
the most part.

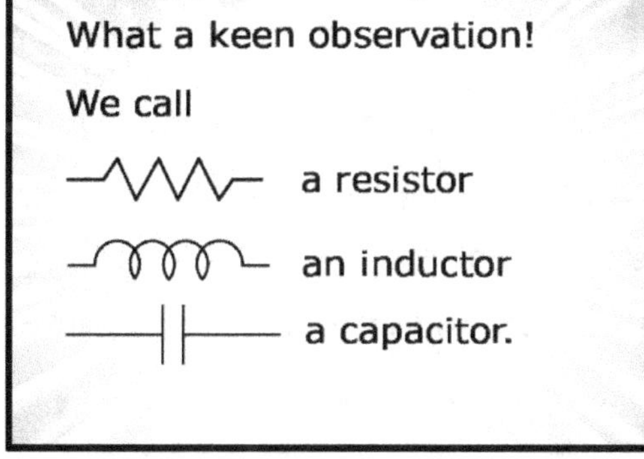

What a keen observation!
We call
a resistor
an inductor
a capacitor.

An electric circuit is
mostly composed of RLCs!

What do the fellows
like resistors, inductors,
and capacitors do?

Resistors are fellows that
impede flows of the current,
inductors are the ones that
impede changes in the current,
and capacitors are the ones that
are like energy tanks that store the current.

Look over there.
Ground water is being pumped up
and sent to the water tank now.

Taking the water flow for the current,
we can say that the rust in the pipe
corresponds to a resistor,
the winding part to an inductor,
and the water tank to a capacitor.

The rust in the pipe plays the role of a resistor since it hampers the flow of the water, and the water tank plays the role of a capacitor since it stores the water, that is, the energy.
I can easily understand that, but I don't know what you mean by the winding part of the pipe.

Let's put the pump in its reverse mode.
It is now harder for the water in the winding part to change the direction of its flow than the one in the straight part, isn't it?

Therefore, we can say that the inductor is like inertia that keeps the flow of water from changing.
inductor = inertia

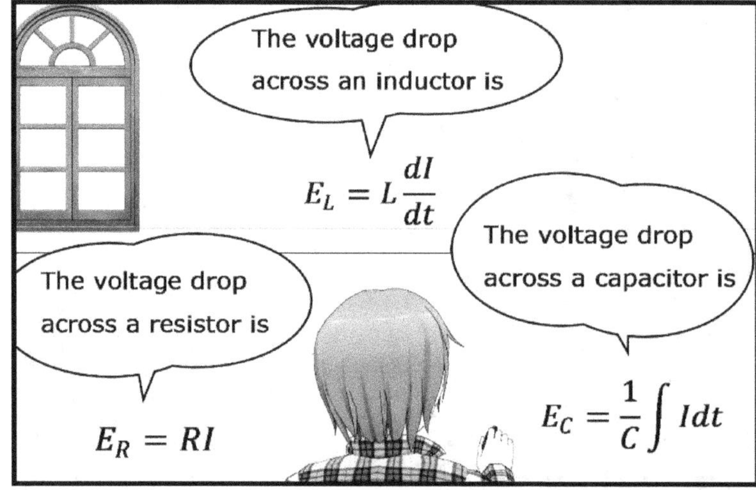

$$E_L = L\frac{dI}{dt}$$

$$E_R = RI$$

$$E_C = \frac{1}{C}\int I\,dt$$

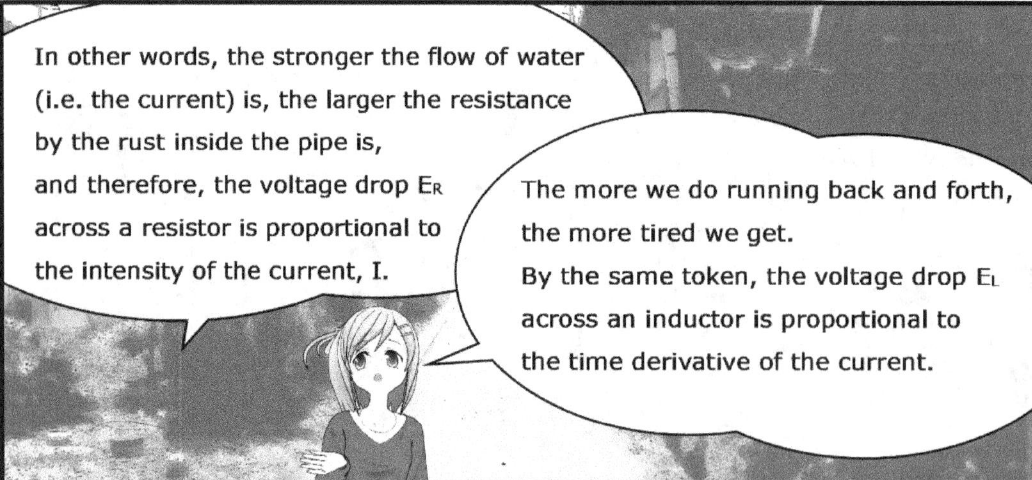

The voltage drop E_C across a capacitor is proportional to $\int I\,dt$ since the amount of the currents stored in a tank, that is, the sum of the currents is denoted by the Σ, which becomes $\int$ if extended.

What are R, L, and C, then?
Those are proportionality constants.

We call
R: Resistance
L: Inductance
C: Capacitance

Why is the C only in its inverse, 1/C, then?
Um, that's because...
We put it that way from the habit.

Have you heard of Kirchhoff's voltage law?
No!

Kirchhoff's voltage law

In a closed loop,

the voltage impressed by the power source

is the same as the sum of the voltage drops

in the rest of the loop.

$$E = E_R + E_L + E_C = RI + L\frac{dI}{dt} + \frac{1}{C}\int I\,dt.$$

DC RL-Circuits

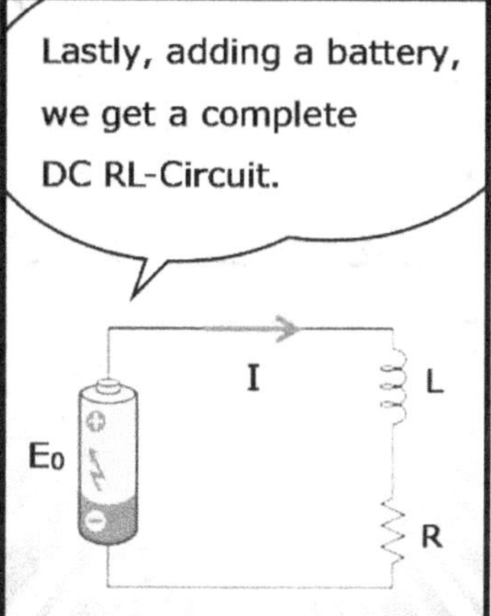

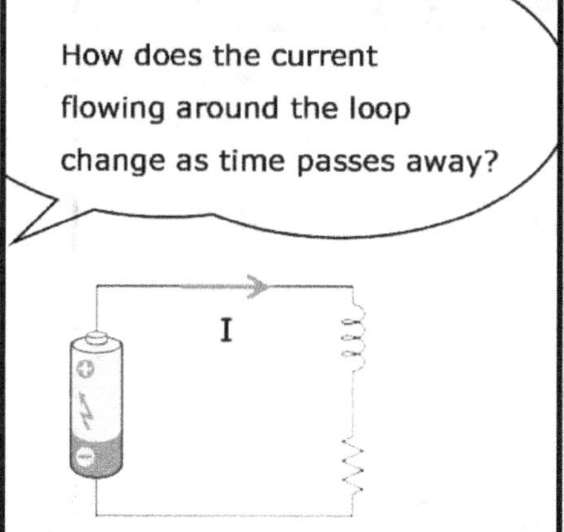

$$E_0 = RI + L\frac{dI}{dt}$$

$$\frac{dI}{dt} + \frac{R}{L}I = \frac{E_0}{L}$$

By the formula, the general solution is

$$y(x) = e^{h}[\int e^{h} r dx + C] \text{ where } h = \int p(x)dx,$$

and if we find the h first, we get

$$h = \int \frac{R}{L} dt = \frac{R}{L} t$$

Therefore

$$I(t) = e^{-\frac{R}{L}t}[\int e^{\frac{R}{L}t} \frac{E_0}{L} dt + C]$$

If we expand what's inside the brackets and rewrite the result, we get

$$I(t) = e^{-\frac{R}{L}t}(\frac{E_0}{L} \frac{L}{R} e^{\frac{R}{L}t} + C)$$

$$= \frac{E_0}{R} + C e^{-\frac{R}{L}t}$$

which is the general solution.

Then, we need to find the constant of integration C by means of the initial condition. Since I = 0 when t = 0 by the initial condition, we get

$$I(0) = \frac{E_0}{R} + C = 0, \qquad C = -\frac{E_0}{R}$$

Therefore, the particular solution satisfying the initial condition is

$$I(t) = \frac{E_0}{R}(1 - e^{-\frac{R}{L}t})$$

$$I(t) = \frac{E_0}{R}\left(1 - e^{-\frac{R}{L}t}\right)$$

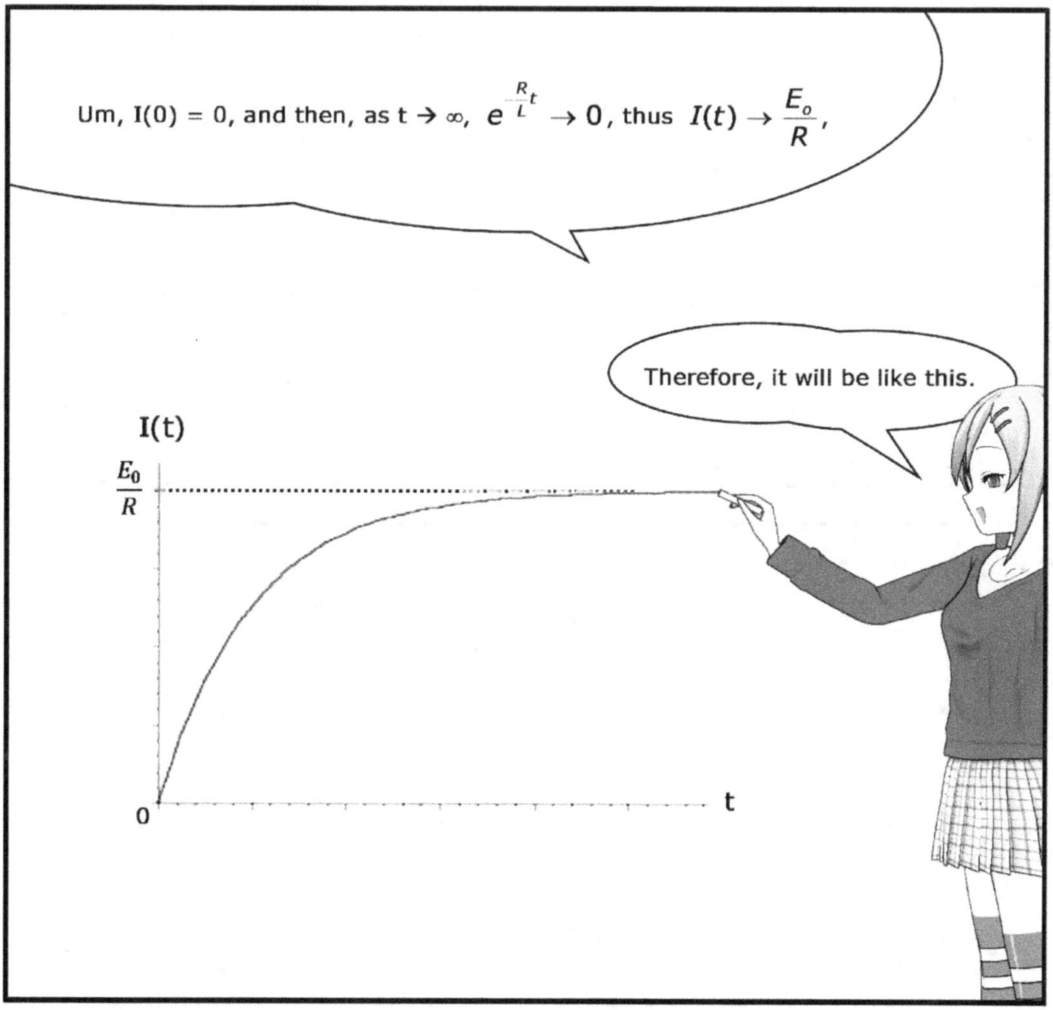

Since the final value is $\dfrac{E_o}{R}$ and it will be 99.9% of the final value, we get

$$I(t) = 0.999 \frac{E_0}{R} = \frac{E_0}{R}\left(1 - e^{-\frac{R}{L}t}\right).$$

Putting the measurements in the solution, we get

$$t = 1.036 \times 10^{-5} \text{ seconds.}$$

If R gets larger,
$e^{-\frac{R}{L}t}$ approaches to 0 quicker
as t increases,
and therefore,
it would be smaller.

What if the inductance L gets larger, then?
L

If L gets larger,
$e^{-\frac{R}{L}t}$ approaches to 0 slower
as t increases,
and therefore,
it would be larger.

You're right, and therefore,
We say that
$\frac{L}{R} = T_L =$ an inductive time constant

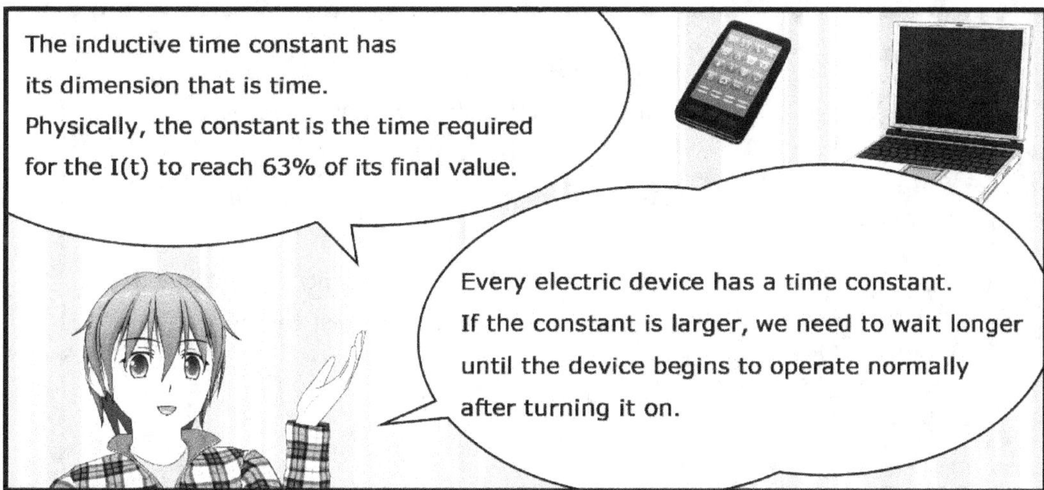

The inductive time constant has
its dimension that is time.
Physically, the constant is the time required
for the I(t) to reach 63% of its final value.
Every electric device has a time constant.
If the constant is larger, we need to wait longer
until the device begins to operate normally
after turning it on.

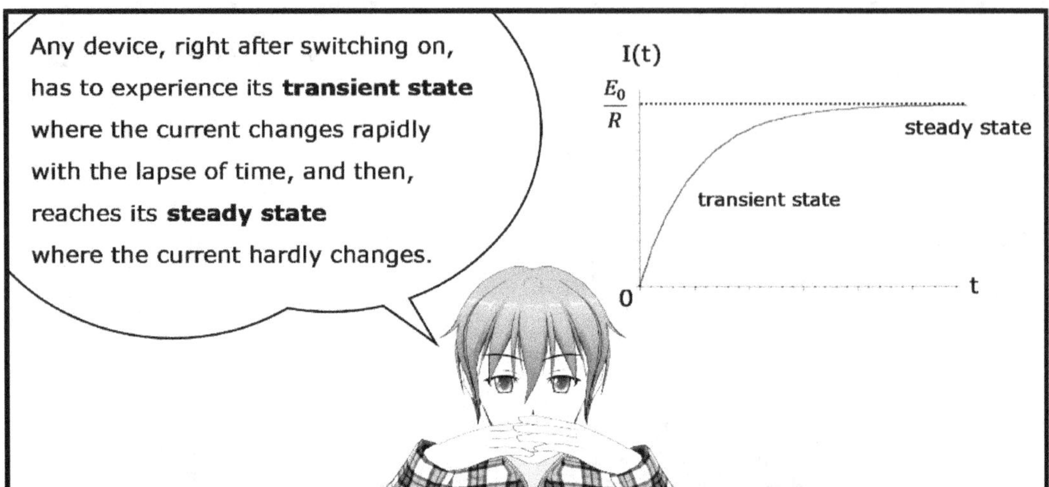

Any device, right after switching on, has to experience its **transient state** where the current changes rapidly with the lapse of time, and then, reaches its **steady state** where the current hardly changes.
I(t)
$\frac{E_0}{R}$
steady state
transient state
0
t

A person also goes through adolescence via childhood, and reaches the steady state after the age of thirty.
Therefore, those in their thirties can hardly change.

Why does the transient state exist?

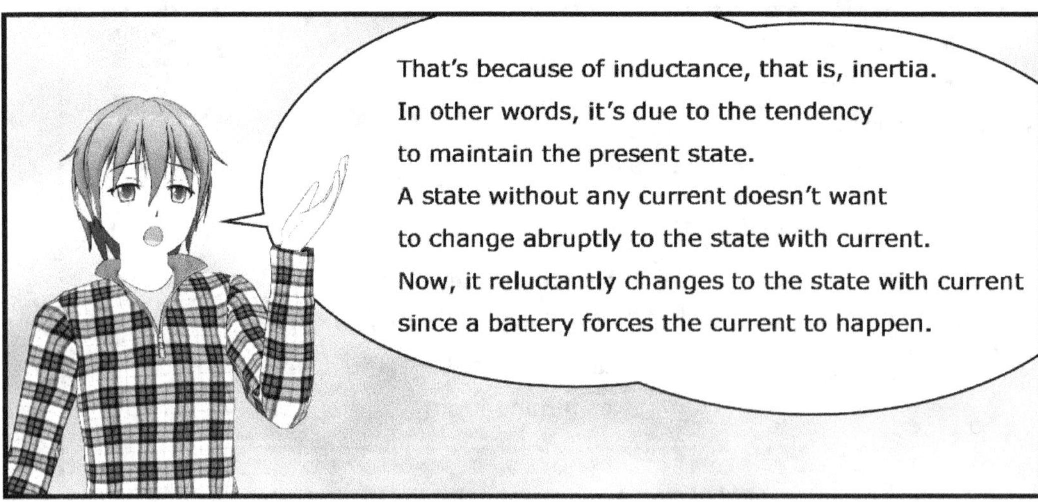

That's because of inductance, that is, inertia.
In other words, it's due to the tendency
to maintain the present state.
A state without any current doesn't want
to change abruptly to the state with current.
Now, it reluctantly changes to the state with current
since a battery forces the current to happen.

Would you change your lazy attitude
to good one at once
if your mom just tells you, "Be diligent"?
If she keeps nagging me
all day everyday,
it might get better bit-by-bit.

The reason that it's hard to change
people's mind is that
they have their own inertias.
My honorable guru,
so much for today!

AC RL-Circuits

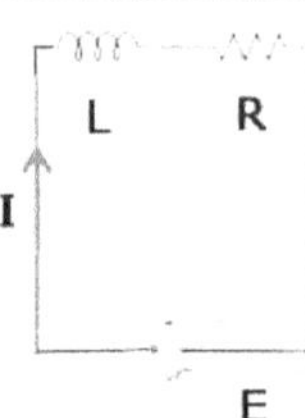

Let's remove the battery from the RL-circuit,
but this time, connect it to the wall plug on this wall.

Then, how does the current on the RL-circuit
change with the laps of time? For simplicity,
let's take the wall plug for the AC source of
$E = E_0 \sin wt$.

Since it is the same RL-circuit, the diff-eq for the current is

$$L\frac{dI}{dt} + RI = E_0 \sin \omega t.$$

Since we need to put the eq in the form of $y' + p(x)y = r(x)$
to make use of the solution formula, we get

$$\frac{dI}{dt} + \frac{R}{L}I = \frac{E_0}{L} \sin \omega t.$$

The solution formula is

$$y(x) = e^{-h}\left[\int e^h r\,dx + C\right], (h = \int p(x)\,dx).$$

If we find the h first

$$h = \int \frac{R}{L}\,dt = \frac{R}{L}t.$$

By substituting h and r,
we get the general solution

$$I(t) = e^{-\frac{R}{L}t}\left[\int e^{\frac{R}{L}t}\frac{E_0}{L}\sin \omega t\,dt + C\right].$$

Except for $\sin \omega t$,
it is exactly the same as
the one in the case of DC!

$$I(t) = Ce^{-\frac{R}{L}t} + \frac{E_0}{R^2 + \omega^2 L^2}(R\sin\omega t - \omega L\cos\omega t).$$

$$I(t) = Ce^{-\frac{R}{L}t} + \frac{E_0}{\sqrt{R^2 + \omega^2 L^2}}\sin(\omega t - \delta)$$

$$\text{where } \delta = \arctan\frac{\omega L}{R}.$$

$$I(0) = C + \frac{E_0}{\sqrt{R^2 + \omega^2 L^2}}(-\sin\delta) = 0$$

$$\therefore C = \frac{E_0}{\sqrt{R^2 + \omega^2 L^2}}\sin\delta.$$

$$I(t) = \frac{E_0}{\sqrt{R^2 + \omega^2 L^2}}\{e^{-\frac{R}{L}t}\sin\delta + \sin(\omega t - \delta)\}.$$

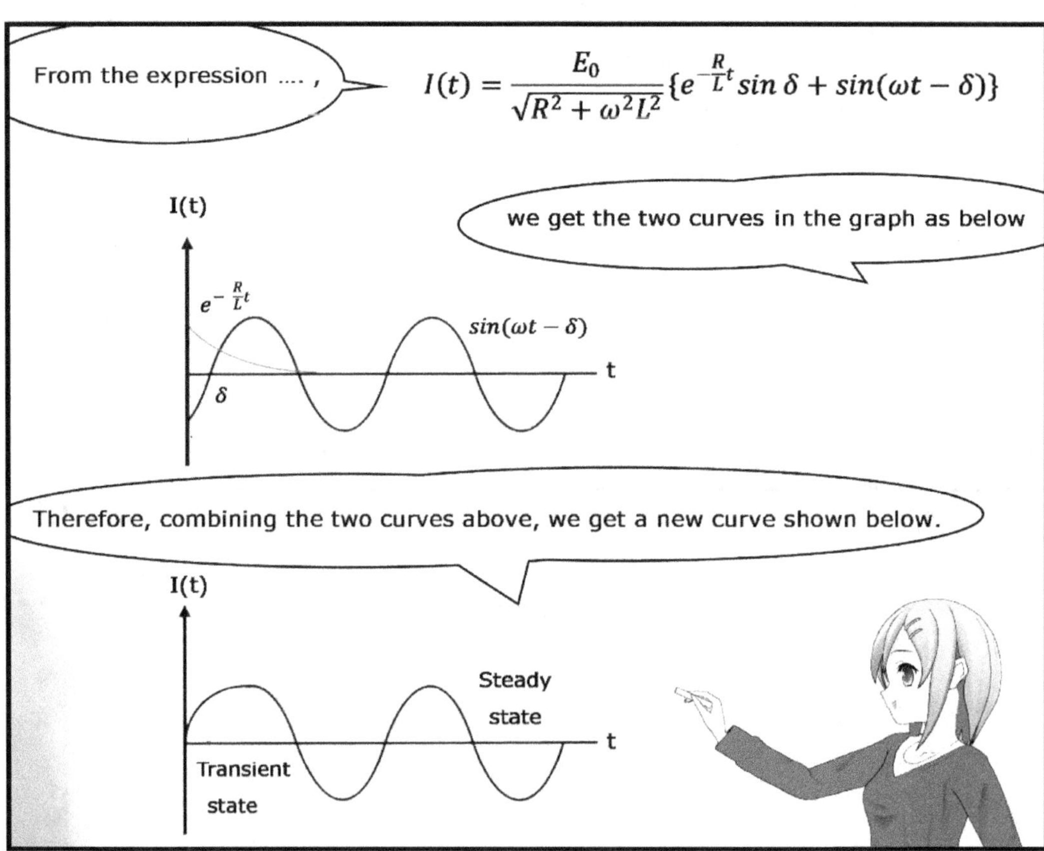

$$I(t) = \frac{E_0}{\sqrt{R^2 + \omega^2 L^2}}\{e^{-\frac{R}{L}t}\sin\delta + \sin(\omega t - \delta)\}$$

Since the input is
AC voltage
in sine function,
it is natural
for the output
to be AC current
in sine function.

In case of AC also,
does an inductance
cause
a transient state?

Yes.
If L → 0, $e^{\frac{R}{L}t}$ → 0
and a transient state
disappears, doesn't it?

Unlike in the case of
DC however,
the inductance affects
steady state also
in the case of AC.
Steady state
also?

The secret
is hidden
in the phase δ.

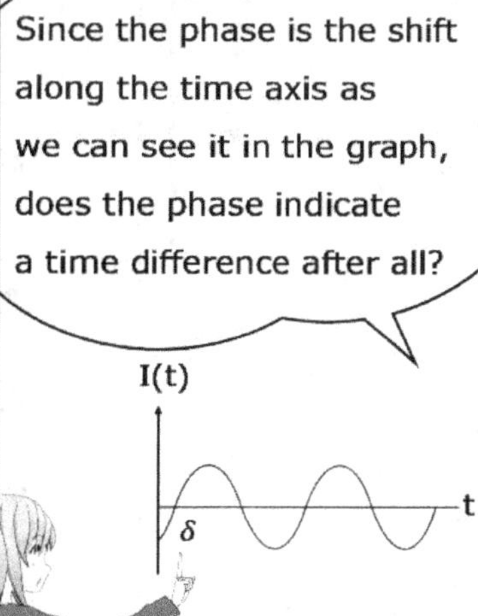

Since the phase is the shift
along the time axis as
we can see it in the graph,
does the phase indicate
a time difference after all?
I(t)
δ
t

That's the stuff!
You're a genius, indeed.
If the inductance L = 0,
what about the phase?

Well, in that case...
If L = 0 in $\delta = \arctan\frac{\omega L}{R}$,
δ = 0 and the phase, that is,
the time difference doesn't exist.

Now, what if $\dfrac{\omega L}{R} \to \infty$?

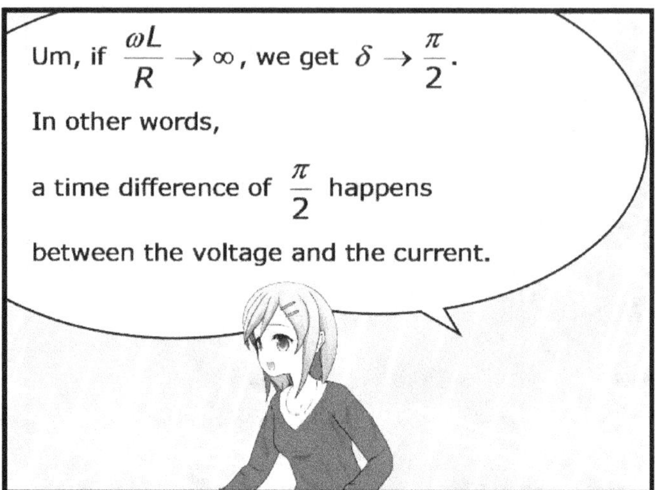

Um, if $\dfrac{\omega L}{R} \to \infty$, we get $\delta \to \dfrac{\pi}{2}$.
In other words,
a time difference of $\dfrac{\pi}{2}$ happens
between the voltage and the current.

Parade rest!

Attention!

Rest! Attention!
Rest! Attention!
R! A! R! A!
AH HAA HAAA

Why come to attention for Parade rest?
By the same token,
if the frequency ω increases,
a time difference of $\dfrac{\pi}{2}$ will happen
between the voltage and the current.
Aha!

DC RC-Circuits

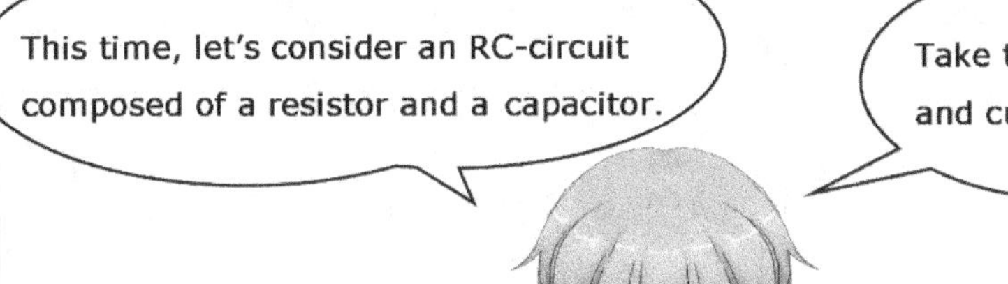

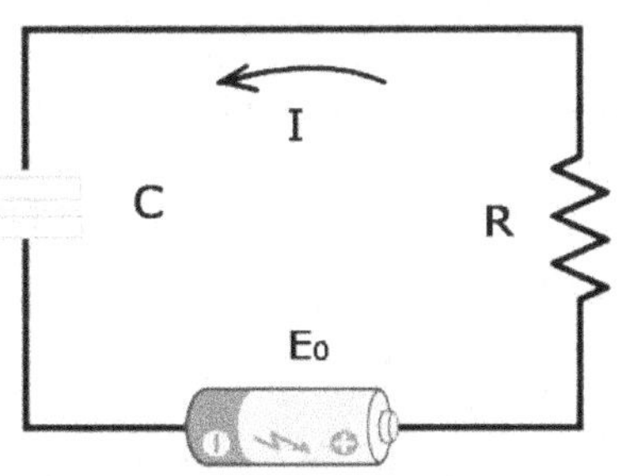

$$RI + \frac{1}{C} \int I\, dt = E_0$$

$$R\frac{dI}{dt} + \frac{1}{C} I = 0$$

$$\frac{dI}{dt} + \frac{1}{RC}I = 0$$

$$y(x) = e^{-h}\left[\int e^h r\, dx + C\right]$$
$$where\ \ h = \int p(x)\, dx$$

$$h = \int \frac{1}{RC}\, dt = \frac{1}{RC}t$$
$$r = 0$$

$$I(t) = \tilde{C}e^{-\frac{t}{RC}}.$$

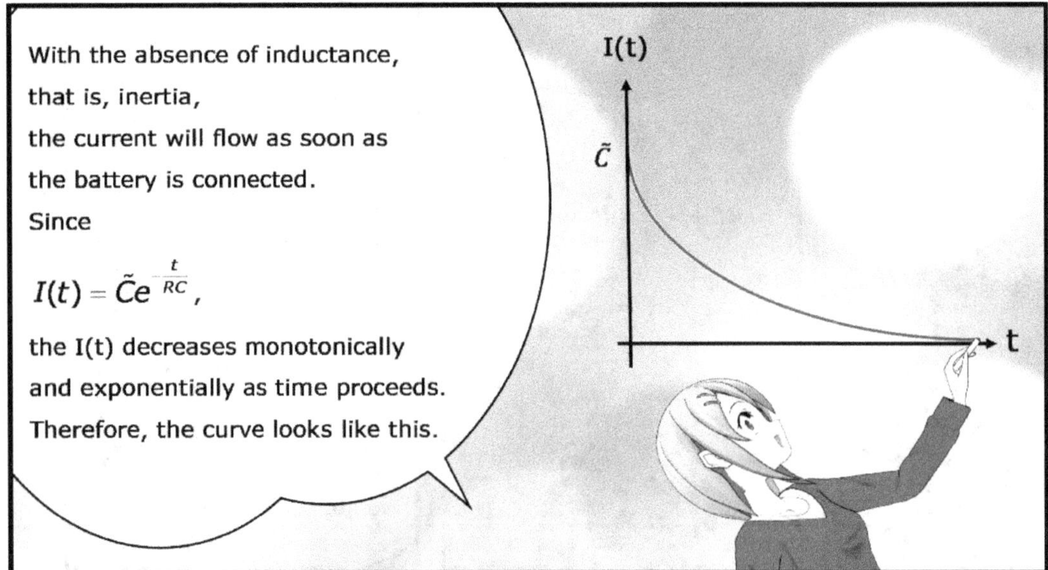

The battery pushes forward charges
inside the wire and
the charges get collected
at the capacitor plates.
When the plates become full of
the charges however,
the battery can no longer
push them forward, and therefore,
the current doesn't flow.

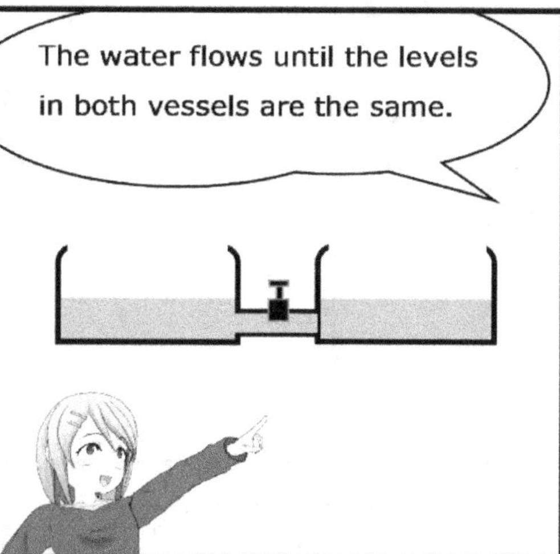

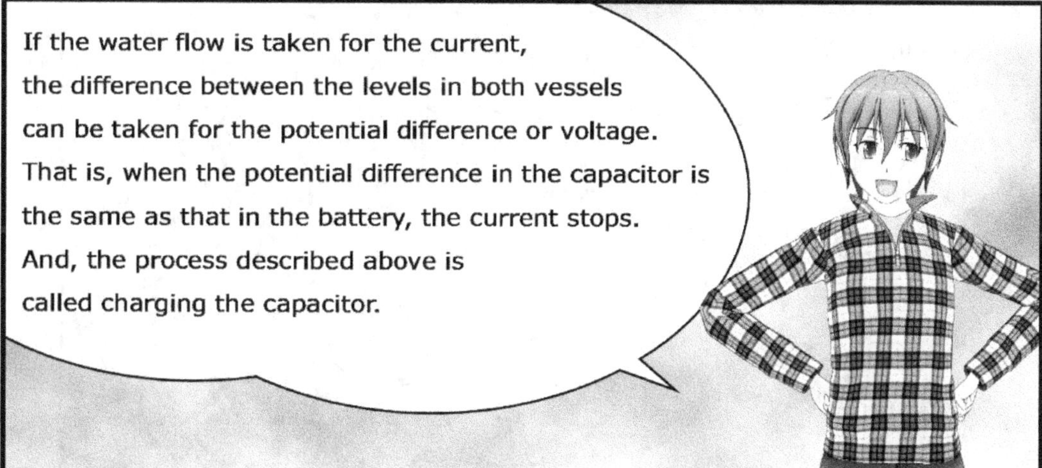

Doesn't an RC-circuit have a time constant?

Yes, it does.
We call RC a capacitive time constant,
and put it in τ_C.

$$\tau_C = RC$$

Then, if R or C is larger,
it takes longer for the current
converges to 0!
You're right.
What was the inductive
time constant?
It was $\tau_L = \dfrac{L}{R}$.

AC RC-Circuits

$$RI + \frac{1}{C} \int I\,dt = E_0 \sin \omega t$$

$$R\frac{dI}{dt} + \frac{1}{C}I = \omega E_0 \cos \omega t$$

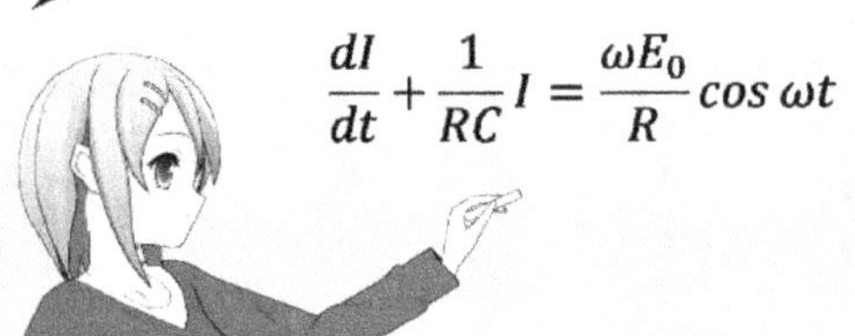

$$\frac{dI}{dt} + \frac{1}{RC}I = \frac{\omega E_0}{R}\cos \omega t$$

$$y(x) = e^{-h}\left[\int e^{h}r\,dx + C\right]$$

$$h = \int \frac{1}{RC}\,dt = \frac{1}{RC}t$$

$$I(t) = e^{-\frac{t}{RC}}\left[\int e^{\frac{t}{RC}}\frac{\omega E_0}{R}\cos \omega t\,dt + \tilde{C}\right]$$

$$= \tilde{C}e^{-\frac{t}{RC}} + \frac{\omega E_0 C}{1+(\omega RC)^2}(\cos \omega t + \omega RC \sin \omega t)$$

$$= \tilde{C}e^{-\frac{t}{RC}} + \frac{\omega E_0 C}{\sqrt{1+(\omega RC)^2}}\sin(\omega t - \delta)$$

$$\left(\delta = \arctan\left(\frac{-1}{\omega RC}\right)\right)$$

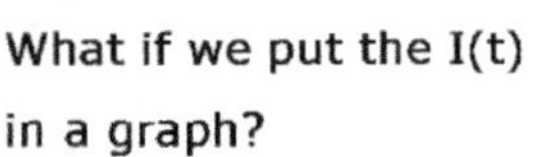

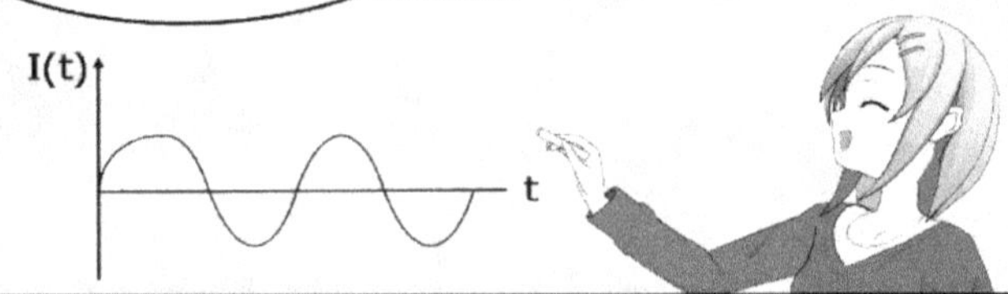

Discharge of a Capacitor

Come on!
You know that the current amounts to
the quantity of charge moved
in one unit of time. So, we have

$$\frac{dQ}{dt} = I, \qquad Q = \int I\, dt$$

See that?

$$RI + \frac{1}{C}\int I\,dt = 0$$

$$R\frac{dQ}{dt} + \frac{Q}{C} = 0$$

$$\frac{dQ}{dt} + \frac{Q}{RC} = 0$$

$$h = \int \frac{1}{RC}\,dt = \frac{1}{RC}t, \quad r = 0$$

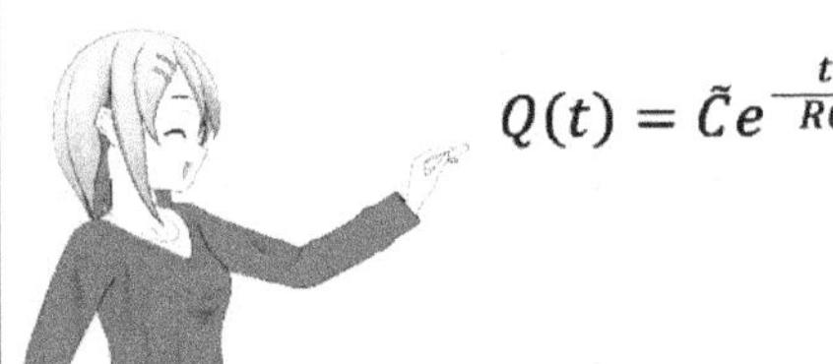

$$y(x) = e^{-h}\left[\int e^{h}\cdot r\,dx + C\right]$$

$$Q(t) = \tilde{C}e^{-\frac{t}{RC}}$$

$$Q(0) = \tilde{C} = Q_0$$

$$Q(t) = Q_0 e^{-\frac{t}{RC}}$$

$$Q(t) = 0.01Q_0 = Q_0 e^{-\frac{t}{RC}}$$

$$t = 4.6RC$$

Attagirl!

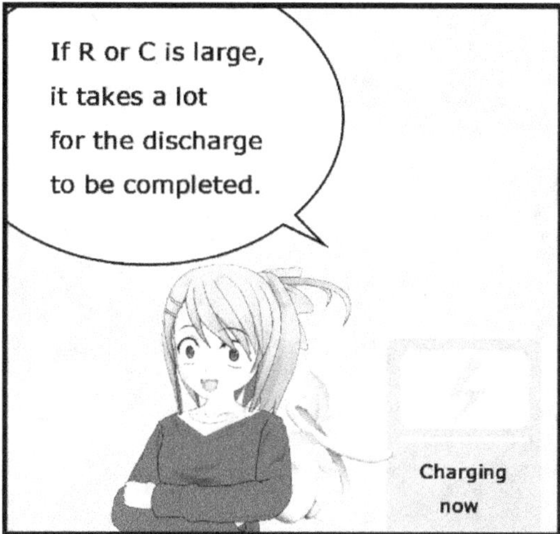

If R or C is large,
it takes a lot
for the discharge
to be completed.
Charging now

It is natural.
If the resistance R is large,
it is harder for the charges to move around.
If the capacitance C is big,
it takes time for all the charges to move out
since the amount stored is large.

Am I snoozing fitfully?
It's exhausting acting
as a guru.

Are we not going to cover the circuit
having all the three,
a resistor, an inductor, and a capacitor?

Sharp as a needle!
In RLC-circuit,
the current has to do with
a 2nd order diff-eq.
that includes second derivatives.
We will cover it tomorrow,
and let's call it a day now.

The essence of math lies entirely in its freedom.
I am Georg Cantor known as the creator of set theory.

Quiz 1

A pacemaker is composed of a battery E_o, a capacitor C, and a switch as shown in the figure below. First, the switch is put to A and the capacitor gets charged, and then, the switch is moved to B, the capacitor gets discharged, and the heart gets an electric stimulus. Repeating such processes, the pacemaker controls the heartbeat by periodic stimuli. Suppose that the switch is at the B, R is the resistance of the heart, and Q_o is the charge stored in the capacitor. Then, set up a diff-eq for the charge Q.

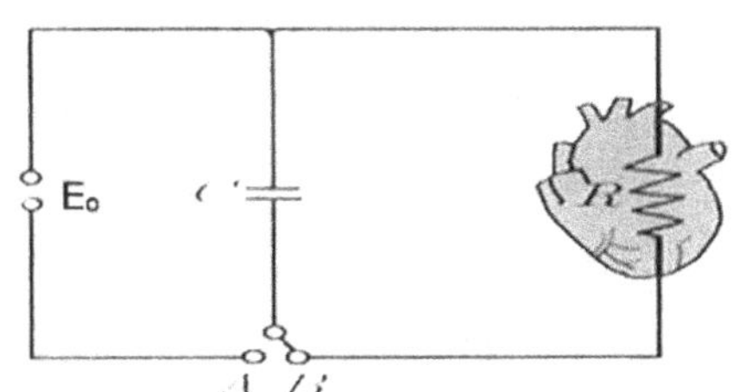

Answer

The Kirchhoff's law says $RI + L\dfrac{dI}{dt} + \dfrac{1}{C}\displaystyle\int I\,dt = E$.

Now, since the closed loop on the right doesn't have an inductor and the power source,

we can see that $L\dfrac{dI}{dt} = 0$, and E = 0. Therefore, $RI + \dfrac{1}{C}\displaystyle\int I\,dt = 0$.

Now, since $I = \dfrac{dQ}{dt}$, we get $\displaystyle\int I\,dt = Q$.

Therefore, the equation is $R\dfrac{dQ}{dt} + \dfrac{Q}{C} = 0$.

The time required for the capacitor to be nearly discharged is t = 4.6RC as calculated earlier. If the same amount of time is required for the capacitor to get charged, the period is T = 2 x 4.6RC at which each electric stimulus is given to the heart. Now, the resistance R in the heart is almost constant.

Consequently, we can calculate and find what capacitor we have to use to provide the heart with the electric stimuli at the period we want.

Quiz 2

What L should we choose in a DC RL-circuit with R= 100 ohm and an initial value I(0)=0 for the current to reach 99.9% of its final value at t=0.01 sec?

Answer

The particular solution satisfying the initial value is

$$I(t) = \frac{E_0}{R}\left(1 - e^{-\frac{R}{L}t}\right). \quad (*)$$

as calculated earlier.

So the final value at $t \to \infty$ $(e^{-\frac{R}{L}t} \to \infty)$ is given by E_0/R.

The 99.9% of the final value is 0.999 E_0/R.

By substituting this into $I(t)$ in eq $(*)$, we get

$$0.999\frac{E_0}{R} = \frac{E_0}{R}(1 - e^{-\frac{R}{L}t})$$

$$0.999 = 1 - e^{-\frac{R}{L}t}$$

$$e^{-\frac{R}{L}t} = 0.001.$$

By taking *In* on both sides, we have

$$\ln e^{-\frac{R}{L}t} = \ln 0.001.$$

Since *In* and *e* are inverse and therefore canceled, we get

$$-\frac{L}{R}t = \ln 0.001. \quad L = -\frac{Rt}{\ln 0.001}$$

By substituting given values, we get

$$L = -\frac{(100\ ohm)(0.01\ s)}{\ln 0.001} = 0.145\ H \quad \text{H: Henry}$$

CHAPTER 5

Since today is the last day, let's visit and have fun at the Science Center.
Whoopee! All right!

Now, here we are!

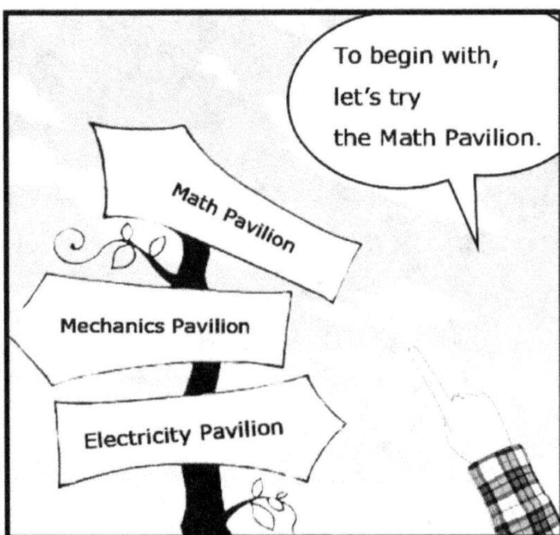

To begin with, let's try the Math Pavilion.
Math Pavilion
Mechanics Pavilion
Electricity Pavilion

1st Order Diff-eq
2nd Order Diff-eq
Higher Order Diff-eq
Velocity
Acceleration
Since we covered 1st order diff-eq on the second day, let's get on the elevator for 2nd order diff-eq.

2nd Order Diff-eq
Now!
Let's get upstairs.
Whiz--

What is this?
Just hit the button.

Whiz--
y"+f(x)+g(x)y=r(x)
Genral Form
Homogeneous

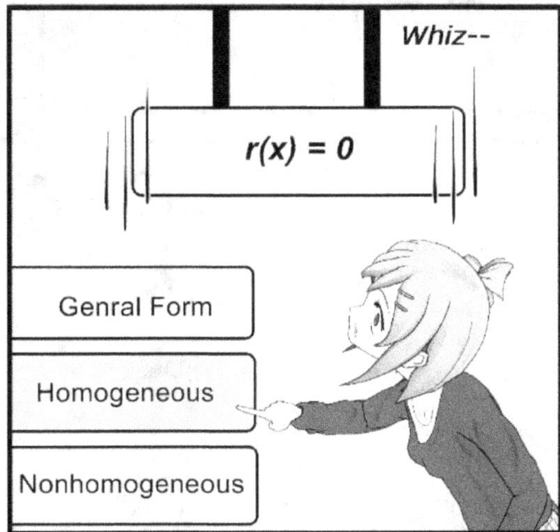
Whiz--
r(x) = 0
Genral Form
Homogeneous
Nonhomogeneous

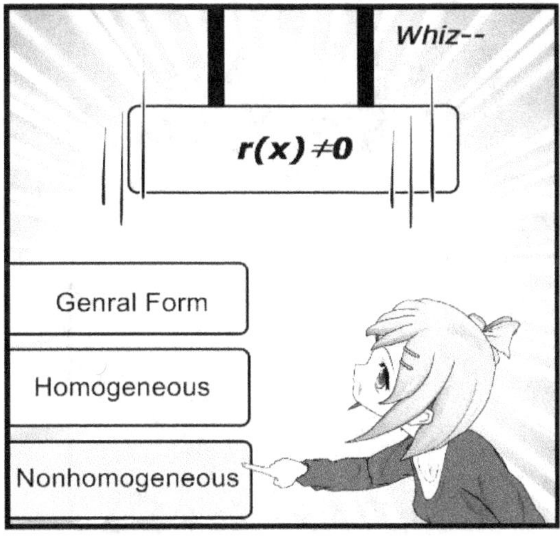
Whiz--
r(x) ≠ 0
Genral Form
Homogeneous
Nonhomogeneous

Here on the second floor,
you will be introduced
the simplest 2nd order diff-eq.
1 2 3

2nd Order homogeneous diff-eq with constant coefficients

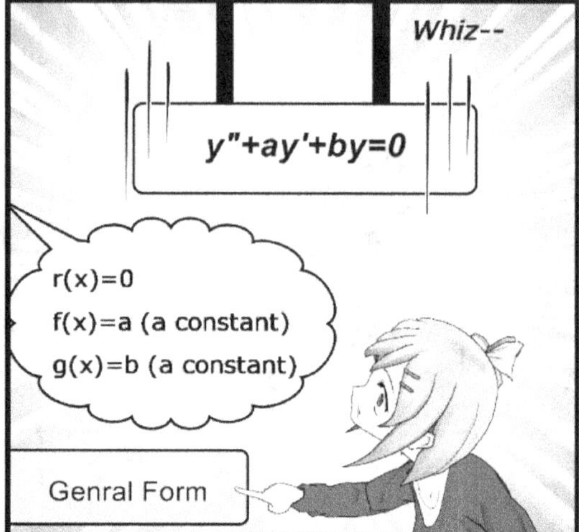

Assuming that $e^{\lambda x}$ is the solution, try it again.

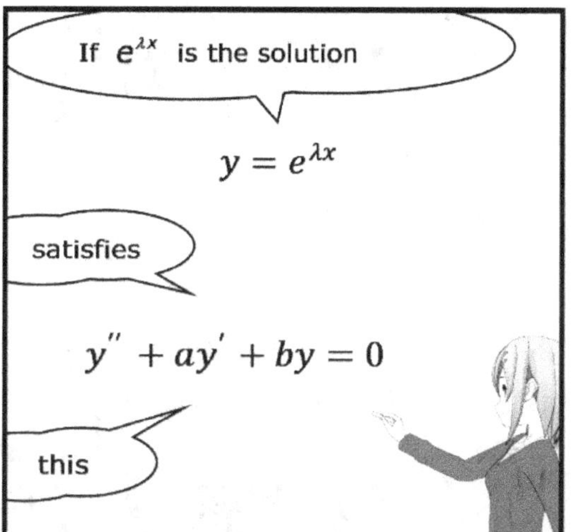

If $e^{\lambda x}$ is the solution
$$y = e^{\lambda x}$$
satisfies
$$y'' + ay' + by = 0$$
this

Finding y″ and y′ for plugging in, we get
$$y' = \lambda e^{\lambda x},$$
$$y'' = \lambda^2 e^{\lambda x}$$

Plugging them in, we get
$$\lambda^2 e^{\lambda x} + a\lambda e^{\lambda x} + b e^{\lambda x} = 0$$
$$(\lambda^2 + a\lambda + b)e^{\lambda x} = 0$$

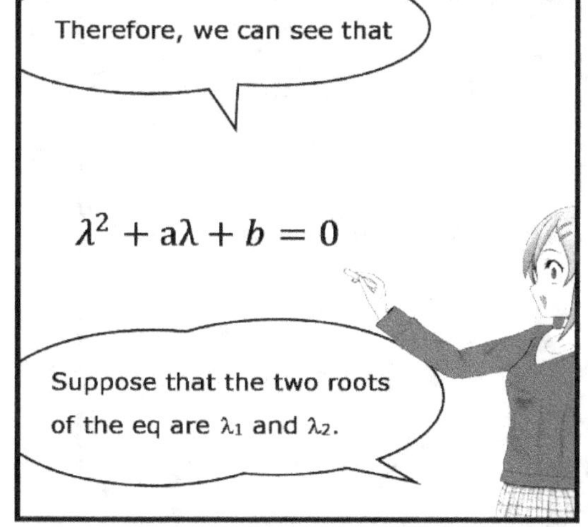

Therefore, we can see that
$$\lambda^2 + a\lambda + b = 0$$
Suppose that the two roots of the eq are λ_1 and λ_2.

Then, the solutions to the diff-eq given are
$$y_1 = e^{\lambda_1 x}, \qquad y_2 = e^{\lambda_2 x}$$

Quiz
How do we call the eq $\lambda^2 + a\lambda + b = 0$?
How do we call it?

Characteristic eq.

What did you say?
Boing!
Ouch --

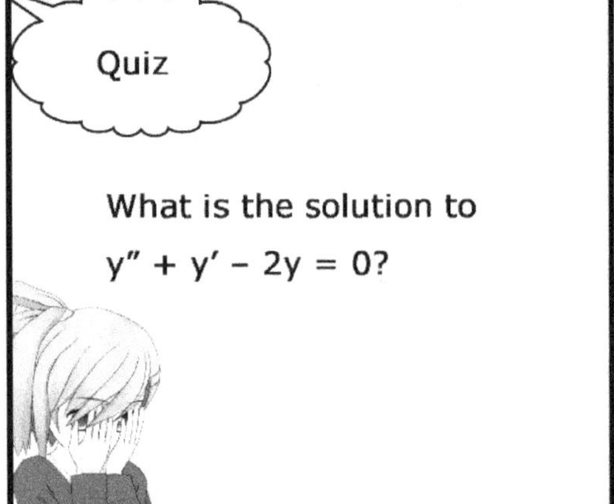

Quiz
What is the solution to $y'' + y' - 2y = 0$?

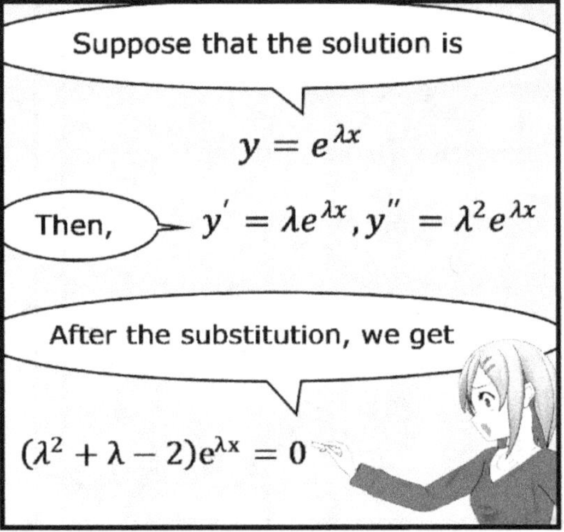

Suppose that the solution is
$y = e^{\lambda x}$
Then, $y' = \lambda e^{\lambda x}, y'' = \lambda^2 e^{\lambda x}$
After the substitution, we get
$(\lambda^2 + \lambda - 2)e^{\lambda x} = 0$

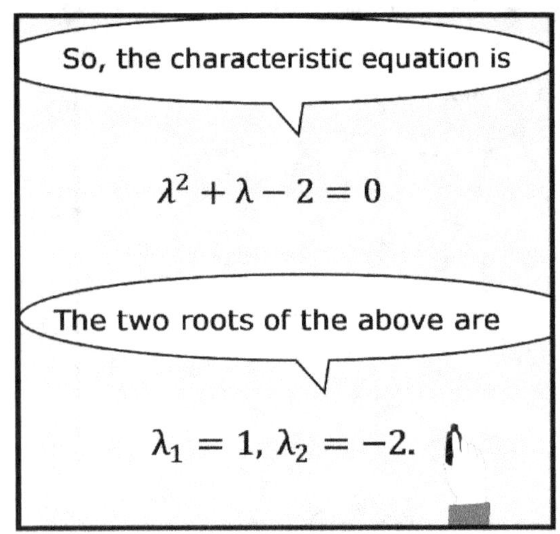

So, the characteristic equation is
$\lambda^2 + \lambda - 2 = 0$
The two roots of the above are
$\lambda_1 = 1, \lambda_2 = -2.$

Therefore, the solutions are
$y_1 = e^x, y_2 = e^{-2x}$

A 2nd order diff-eq has two solutions also as a quadratic eq does.
Boing!
Ouch --

It looks like it, doesn't it? However, a careful thought will let you see it is not the case.

For instance, is 2e^x the solution to y'' + y' − 2y = 0 or not?
Plugging it in, we'll see if it is.

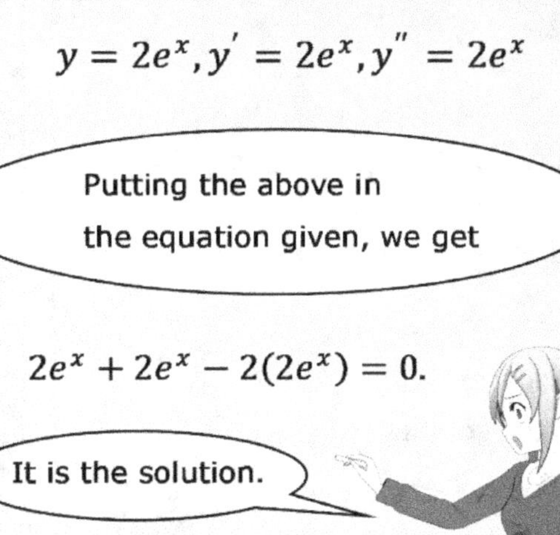
$y = 2e^x, y' = 2e^x, y'' = 2e^x$
Putting the above in the equation given, we get
$2e^x + 2e^x - 2(2e^x) = 0.$
It is the solution.

Is 5e^x a solution or not, then?
I guess it is.

What about Ce^x where C is constant, then?
It is, I guess!

How about
e^x + e^{-2x},
then?
Putting it in,
I will see if it is.

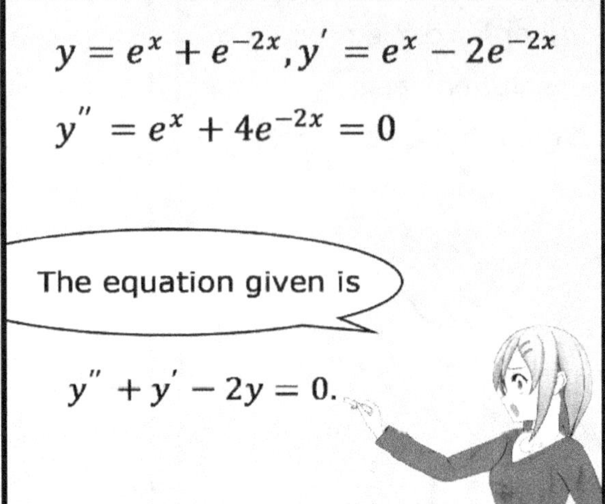
$y = e^x + e^{-2x}, y' = e^x - 2e^{-2x}$
$y'' = e^x + 4e^{-2x} = 0$
The equation given is
$y'' + y' - 2y = 0.$

Then,
$(e^x + 4e^{-2x}) + (e^x - 2e^{-2x})$
$-2(e^x + e^{-2x}) = 0.$
Therefore,
it is the solution.

How about
2e^x + 3e^{-2x}, then?
I guess.

It is the case for
homogeneous eq only
that the sum of
two solutions can be
a solution also.
Should I let her know
that or not?
How many solutions
are there, then?
I can see
infinitely many

What if we put
in an expression
all such infinitely
many solutions?

$y = C_1 e^x + C_2 e^{-2x}$
where C_1 and C_2 are constants
Here it is!

Quiz
What is the general solution to $y'' + y' - 2y = 0$?
$y = C_1 e^x + C_2 e^{-2x}$

Quiz
What are the bases of the solutions to $y'' + y' - 2y = 0$?
They are e^x and e^{-2x}.
Bases?

Aha! It is the case not that the solutions are two but that the bases are two.

Quiz
After all, solving 2nd order diff-eq, what do we do?

We find two bases, multiply each by a constant, and add the products, and then, take the sum for the solution.
Whew!

Now, can we use e^x and $2e^x$ for the bases instead of the bases e^x and e^{-2x}?
Shush!
Whop!
Ouch!

We can't, but let's see why not. If that were the case, what would be the solution?
It would be this.
$y = C_1 e^x + 2C_2 e^x$.

It'll be eventually
$y = (C_1 + 2C_2)e^x$
$= Ce^x$.
Therefore
The solution has one basis only cannot be the general solution representing the infinitely many solutions. Therefore, we can't do that.

To get the general solution, we need to use two bases linearly independent.
Linearly independent!

Yipes!
The floor is moving.

You are
in the second room
now.
What's in the box?
Squeak!

I can see
apples and pears.

Right on, that's the right answer.
What if I reply that there are
red apples and green ones?

The answer is not wrong, and yet is stupid.
That's because the answer is missing
the significant fact that fruits in another kind,
pears are included.

Exactly!
In this case,
an apple and a pear
are taken for the bases.

Now,
adding or cutting apples
doesn't make a pear.
Then, apples and pears
are said to be
linearly independent.

On the other hand,
red apples and green apples
are said to be linearly dependent.
Removing the adjectives in front,
we get just apples.

For instance, e^x and $2e^x$
are linearly dependent.

If y_1 and y_2 are linearly dependent,
$y_1 = ky_2$ (k: constant).

The floor is moving again!
Yippee! Fireworks!
Review Session
This is the last room to visit.

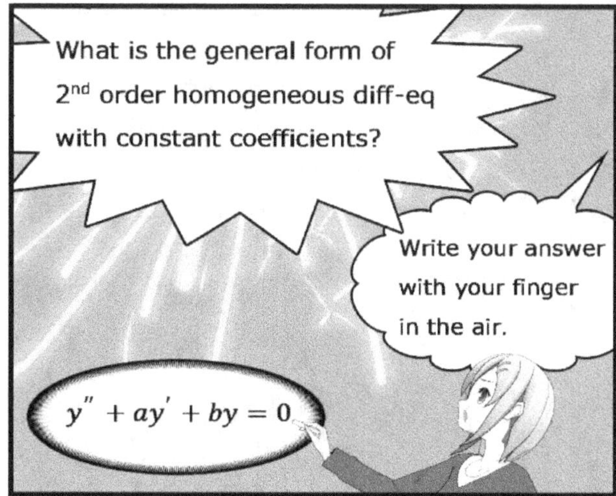

What is the general form of 2nd order homogeneous diff-eq with constant coefficients?
Write your answer with your finger in the air.
$y'' + ay' + by = 0$

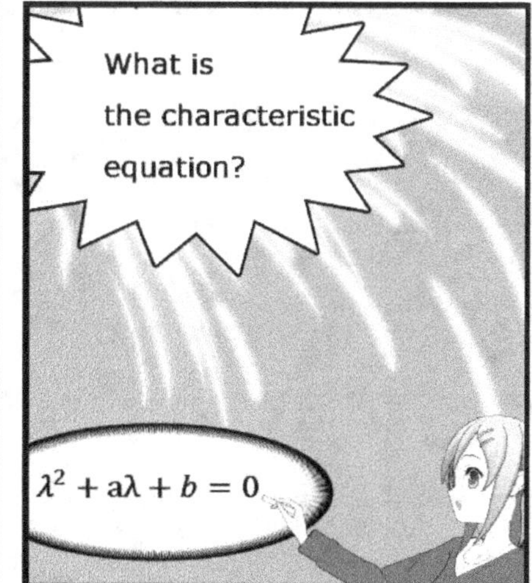

What is the characteristic equation?
$\lambda^2 + a\lambda + b = 0$

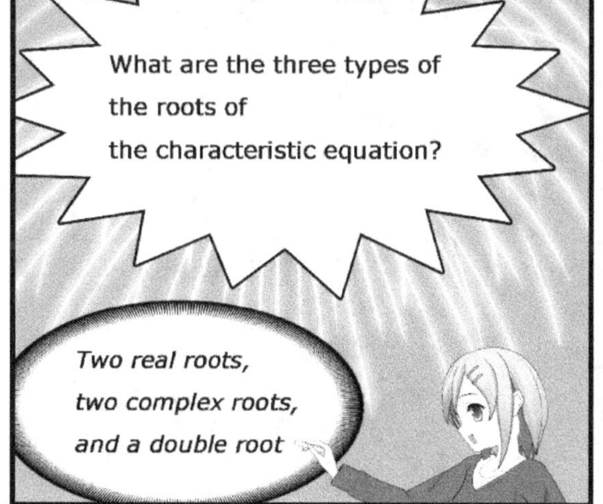

What are the three types of the roots of the characteristic equation?
Two real roots, two complex roots, and a double root

If the characteristic equation has two real roots λ_1 and λ_2, what is the general solution?

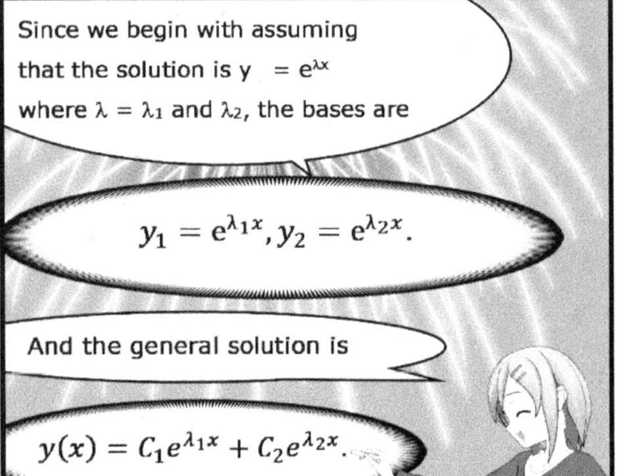

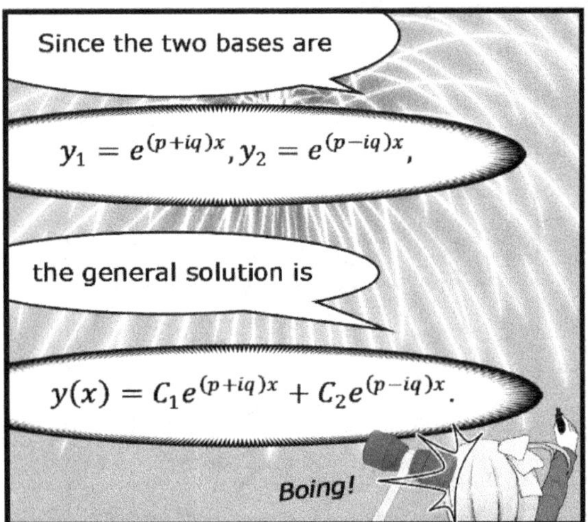

Since a physical quantity measurable is in real number, we need to find the bases in real number.
Try the Euler's formula
$e^{\pm i\theta} = \cos\theta \pm i\sin\theta.$

$$y_1 = e^{(p+iq)x} = e^{px} e^{iqx} = e^{px}(\cos qx + i\sin qx)$$

$$y_2 = e^{(p-iq)x} = e^{px} e^{-iqx} = e^{px}(\cos qx - i\sin qx)$$

Eliminating the i to make real bases, we get

These two bases are linearly independent. Therefore, the general solution is

$$\frac{1}{2}(y_1 + y_2) = e^{px}\cos qx, \quad \frac{1}{2i}(y_1 - y_2) = e^{px}\sin qx$$

$$y(x) = e^{px}(C_1 \cos qx + C_2 \sin qx).$$

What is the general solution when the characteristic equation has a double root?

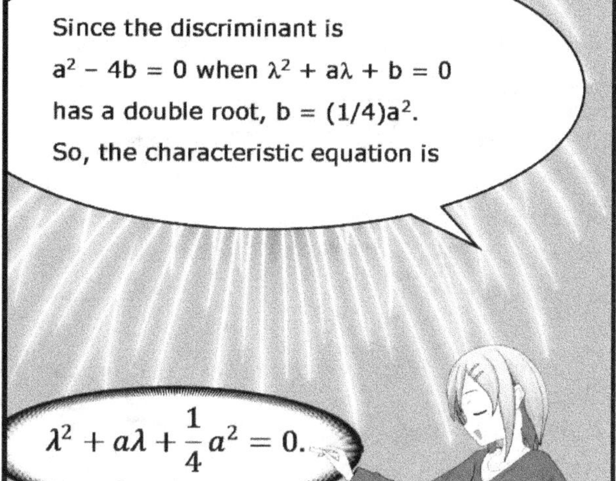

Since the discriminant is
$a^2 - 4b = 0$ when $\lambda^2 + a\lambda + b = 0$
has a double root, $b = (1/4)a^2$.
So, the characteristic equation is
$\lambda^2 + a\lambda + \frac{1}{4}a^2 = 0.$

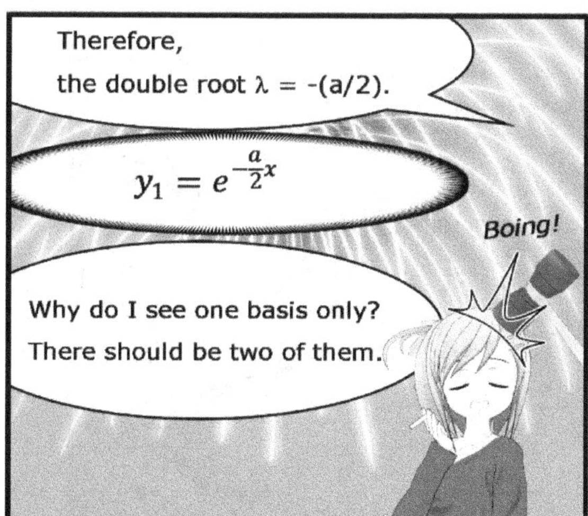

Therefore,
the double root $\lambda = -(a/2)$.
$y_1 = e^{-\frac{a}{2}x}$
Boing!
Why do I see one basis only?
There should be two of them.

Assume that the other basis is
$y_2(x) = u(x)y_1(x).$
Find the u first, and then, find the y_2.

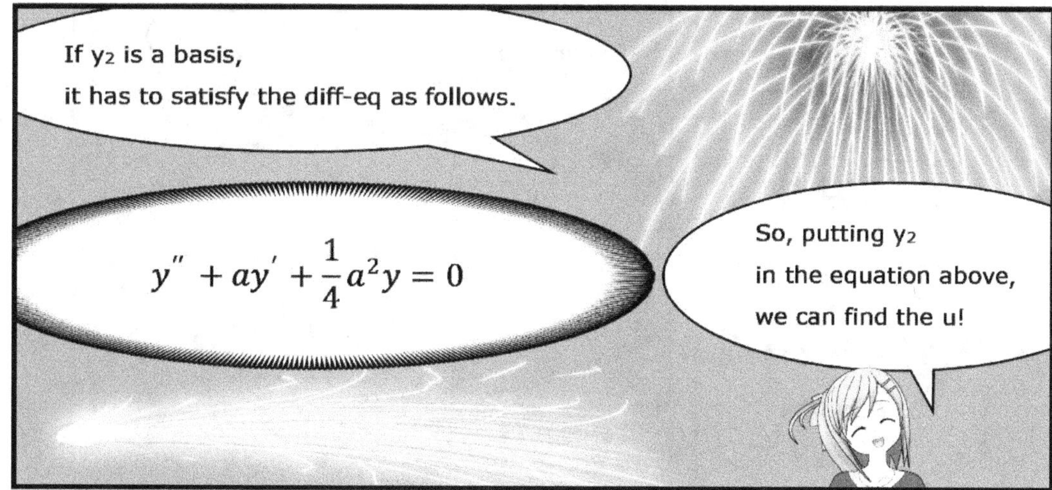

If y_2 is a basis,
it has to satisfy the diff-eq as follows.
$y'' + ay' + \frac{1}{4}a^2 y = 0$
So, putting y_2
in the equation above,
we can find the u!

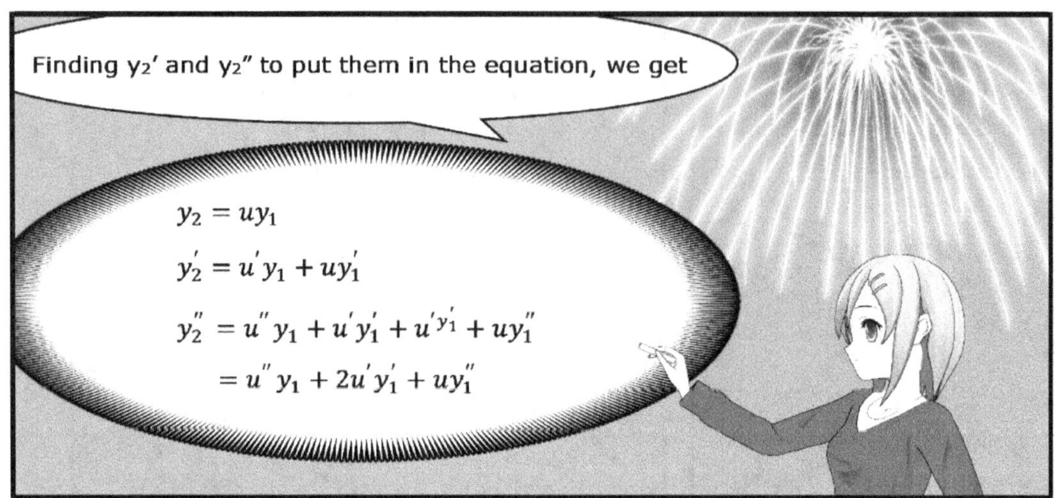

$$y_2 = uy_1$$
$$y_2' = u'y_1 + uy_1'$$
$$y_2'' = u''y_1 + u'y_1' + u'y_1' + uy_1''$$
$$= u''y_1 + 2u'y_1' + uy_1''$$

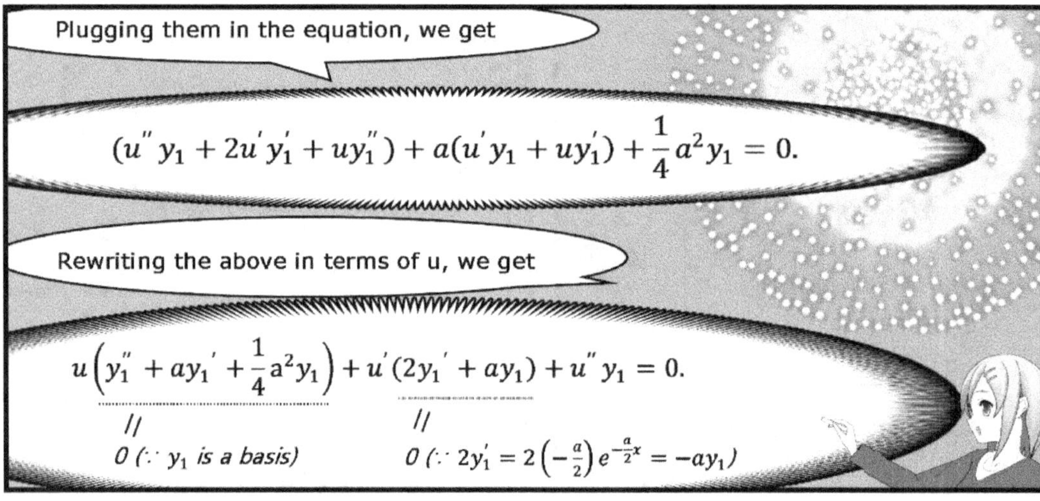

$$(u''y_1 + 2u'y_1' + uy_1'') + a(u'y_1 + uy_1') + \frac{1}{4}a^2y_1 = 0.$$

$$u\left(y_1'' + ay_1' + \frac{1}{4}a^2y_1\right) + u'(2y_1' + ay_1) + u''y_1 = 0.$$

$$\underbrace{}_{0\ (\because\ y_1\ is\ a\ basis)} \qquad \underbrace{}_{0\ \left(\because\ 2y_1' = 2\left(-\frac{a}{2}\right)e^{-\frac{a}{2}x} = -ay_1\right)}$$

2nd Order Homogeneous Diff-eq with Constant Coefficients

The general form $\qquad y'' + ay' + by = r(x)$

The characteristic equation $\qquad \lambda^2 + a\lambda + b = 0$

The general solution

Two real roots (λ_1, λ_2) $\qquad y(x) = C_1 e^{\lambda_1 x} + C_2 e^{\lambda_2 x}$

Two complex roots $(\lambda_{1,2} = p \pm iq)$ $\qquad y(x) = e^{px}(C_1 \cos qx + C_2 \sin qx)$

A double root (λ) $\qquad y(x) = (C_1 + C_2 x)e^{\lambda x}$

The signs in math are no more than
symbols to be performed.
However, they are for doing
hand movements mechanically repeated
since they reduce the burdens
on our brains and help save brainpower
for much more important and difficult functions.
Have you heard of
Mach numbers,
Mach velocities,
Mach principles,
Mach bands,
Mach effects, etc.?
My name keeps
repeating in the question.
I am Ernst Mach.

Quiz 1   Find the general solution to $y'' - y = 0$.

Answer   Assuming that the basis is $y = e^{\lambda x}$, and finding y'', we get

$$y' = \lambda e^{\lambda x}, \quad y'' = \lambda^2 e^{\lambda x}.$$

Substituting it to the given eq, we get

$$\lambda^2 e^{\lambda x} - e^{\lambda x} = 0, \quad (\lambda^2 - 1)e^{\lambda x} = 0.$$

So, the characteristic eq is $\lambda^2 - 1 = 0$, and in turn, $\lambda = \pm 1$. (two real roots)

Therefore, the bases are e^x and e^{-x}.

The general solution is $y = C_1 e^x + C_2 e^{-x}$.

Quiz 2   Find the general solution to $y'' + 8y' + 16y = 0$.

Answer   The characteristic eq is $\lambda^2 + 8\lambda + 16 = 0$, and then, $\lambda = -4$. (a double root)

Therefore, the bases are e^{-4x} and xe^{-4x}.

The general solution is $y = (C_1 + C_2 x)e^{-x}$.

Quiz 3   Find the general solution to $y'' - 2y' + 10y = 0$.

Answer   The characteristic eq is $\lambda^2 - 2\lambda + 10 = 0$,

and then, $\lambda = 1 \pm 3i$. (two complex roots)

Therefore, the bases are $e^x \cos 3x$ and $e^x \sin 3x$.

The general solution is $y = e^x(C_1 \cos 3x + C_2 \sin 3x)$.

Quiz 4

We have $y'' + 2y' + 5y = 0$. Then:

(a) Find the general solution.

(b) Assuming that the initial values are $y(0) = 1$ and $y'(0) = 5$, find the particular solution.

Answer

(a) The characteristic eq is $\lambda^2 + 2\lambda + 5 = 0$,

and then, $\lambda = -1 \pm 2i$ (two complex roots)

Therefore, the bases are $e^{-x}\cos 2x$ and $e^{-x}\sin 2x$.

The general solution is $y = e^{-x}(C_1\cos 2x + C_2\sin 2x)$.

(b) We can find the constants C_1 and C_2 using the initial condition.

$$y(0) = e^{-(0)}(C_1\cos(2 \times 0) + C_2\sin(2 \times 0)) = C_1 = 1$$
$$y' = -e^{-x}(C_1\cos 2x + C_2\sin 2x) + e^{-x}(-2C_1\sin 2x + 2C_2\cos 2x)$$
$$y'(0) = -C_1 + 2C_2 = 5$$

Putting 1 into the C_1 → $C_2 = 3$.

Therefore, the particular solution satisfying

the initial condition is $y = e^{-x}(\cos 2x + 3\sin 2x)$.

2nd Order Nonhomogeneous Diff-eq with Constant Coefficients

Whiz--
Prove that the general solution to a nonhomogeneous diff-eq is
y(x) = yh(x) + yp(x)
where yh(x) is the general solution to the homogeneous diff-eq
and yp(x) is the particular solution to the nonhomogeneous diff-eq.

We need to show that
y = yh + yp
satisfies
y"+ay'+by = r.

Finding y" and y' to substitute with $y'' = y_h'' + y_p''$, $y' = y_h' + y_p'$.

Hurrah!
After the substitutions, we get
$$\left(y_h'' + y_p''\right) + a\left(y_h' + y_p'\right) + b\left(y_h + y_p\right)$$
$$= (y_h'' + ay_h' + by_h) + \left(y_p'' + ay_p' + by_p\right) = r.$$
‖   ‖
0   r
(Since yh is a solution   (Since yp is a solution
to the homogeneous eq.)   to the nonhomogeneous eq.)
Q.E.D.

Finding the y_h, we can follow the method covered
on the second floor,
but what about finding the y_p?

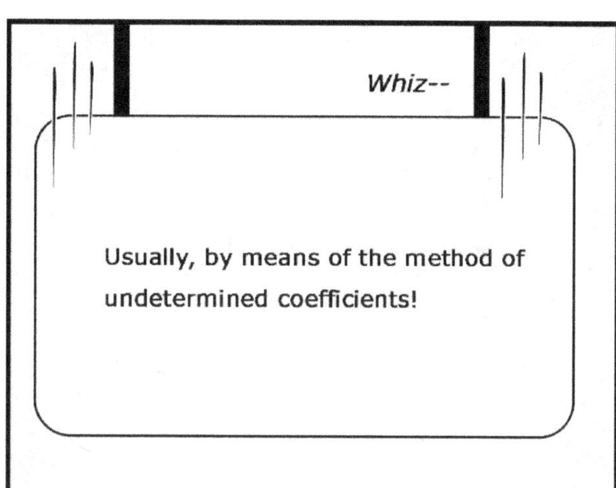
Whiz--
Usually, by means of the method of undetermined coefficients!

What is the method of undetermined coefficients?

The method of undetermined coefficients is the method
where assuming that y_p is in the similar form of r(x),
we assign some appropriate coefficients to the terms in y_p,
and then,
determine the coefficients so that the y_p can satisfy the equation given.

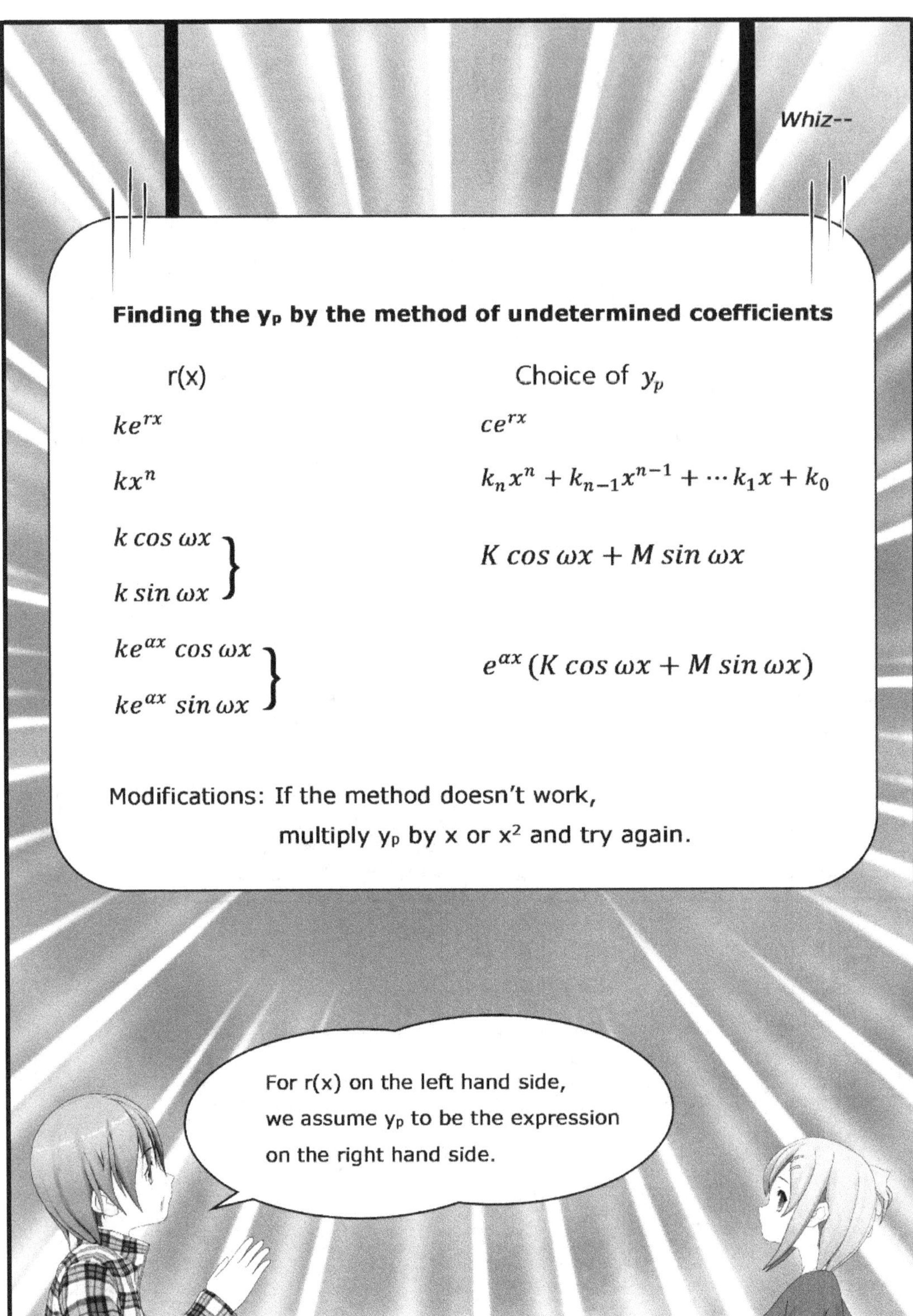

Finding the y_p by the method of undetermined coefficients

r(x)	Choice of y_p
ke^{rx}	ce^{rx}
kx^n	$k_n x^n + k_{n-1} x^{n-1} + \cdots k_1 x + k_0$
$k \cos \omega x$ $k \sin \omega x$	$K \cos \omega x + M \sin \omega x$
$ke^{\alpha x} \cos \omega x$ $ke^{\alpha x} \sin \omega x$	$e^{\alpha x} (K \cos \omega x + M \sin \omega x)$

Modifications: If the method doesn't work,

multiply y_p by x or x^2 and try again.

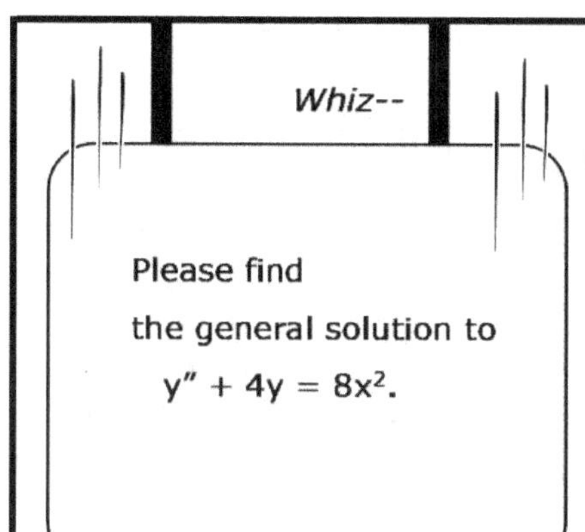

1) Finding the y_h:

The y_h is the general solution to
$$y'' + 4y = 0.$$
The characteristic equation is
$$\lambda^2 + 4 = 0.$$
So $\lambda_1 = 2i$, $\lambda_2 = -2i$ Therefore
$$y_h = C_1 \cos 2x + C_2 \sin 2x.$$

2) Finding the y_p:

Since $r(x) = 8x^2$

We assume $y_p = K_2 x^2 + K_1 x + K_0$

Put y_p into $y'' + 4y = 8x^2$. Then, we get
$$2K_2 + 4(K_2 x^2 + K_1 x + K_0) = 8x^2. \quad (\because\ y_p'' = 2K_2)$$

Rewriting the above in terms of x, we get
$$4K_2 x^2 + 4K_1 x + (2K_2 + 4K_0) = 8x^2.$$

Consequently
$$4K_2 = 8, \qquad 4K_1 = 0, \qquad 2K_2 + 4K_0 = 0$$
$$K_2 = 2, \qquad K_1 = 0, \qquad K_0 = -1$$

So
$$y_p = 2x^2 - 1.$$

Therefore
$$y = y_h + y_p$$
$$= C_1 \cos 2x + C_2 \sin 2x + 2x^2 - 1$$

Quiz

Find the general solution to $y'' - 4y' + 3y = 10e^{-2x}$.

Answer

The general solution to the nonhomogeneous diff-eq is
$y(x) = y_h(x) + y_p(x)$ where y_h is the particular solution
to the nonhomogeneous diff-eq
and y_p is the general solution to the homogeneous diff-eq.

Step 1: Finding the y_h

The characteristic eq of $y'' - 4y + 3y = 0$ is $\lambda^2 - 4\lambda + 3 = 0$.
So, $\lambda = 1$ or 3.
Therefore, $y_h = C_1e^x + C_2e^{3x}$.

Step 2: Finding the y_p
Assume $y_p = Ce^{-2x}$ since the right hand side is $10e^{-2x}$.
(Method of Undetermined Coefficients)
Put into $y'' - 4y' + 3y = 10e^{-2x}$ the $y_p' = -2Ce^{-2x}$ and the $y_p'' = 4Ce^{-2x}$.

$$4Ce^{-2x} - 4(-2Ce^{-2x}) + 3(Ce^{-2x}) = 10e^{-2x}$$

So
$$4C+8C+3C=10, \quad C=2/3.$$

Therefore
$$y_p = \frac{2}{3}e^{-2x}$$

Step 3: Finding the $y(x)$

$$y(x) = y_h + y_p = C_1e^x + C_2e^{3x} + \frac{2}{3}e^{-2x}$$

2nd Order Diff-eq where coefficients are not constant

$$\sum_{n=0}^{\infty} a_n x^n = a_0 + a_1 x + a_2 x^2 + \cdots$$

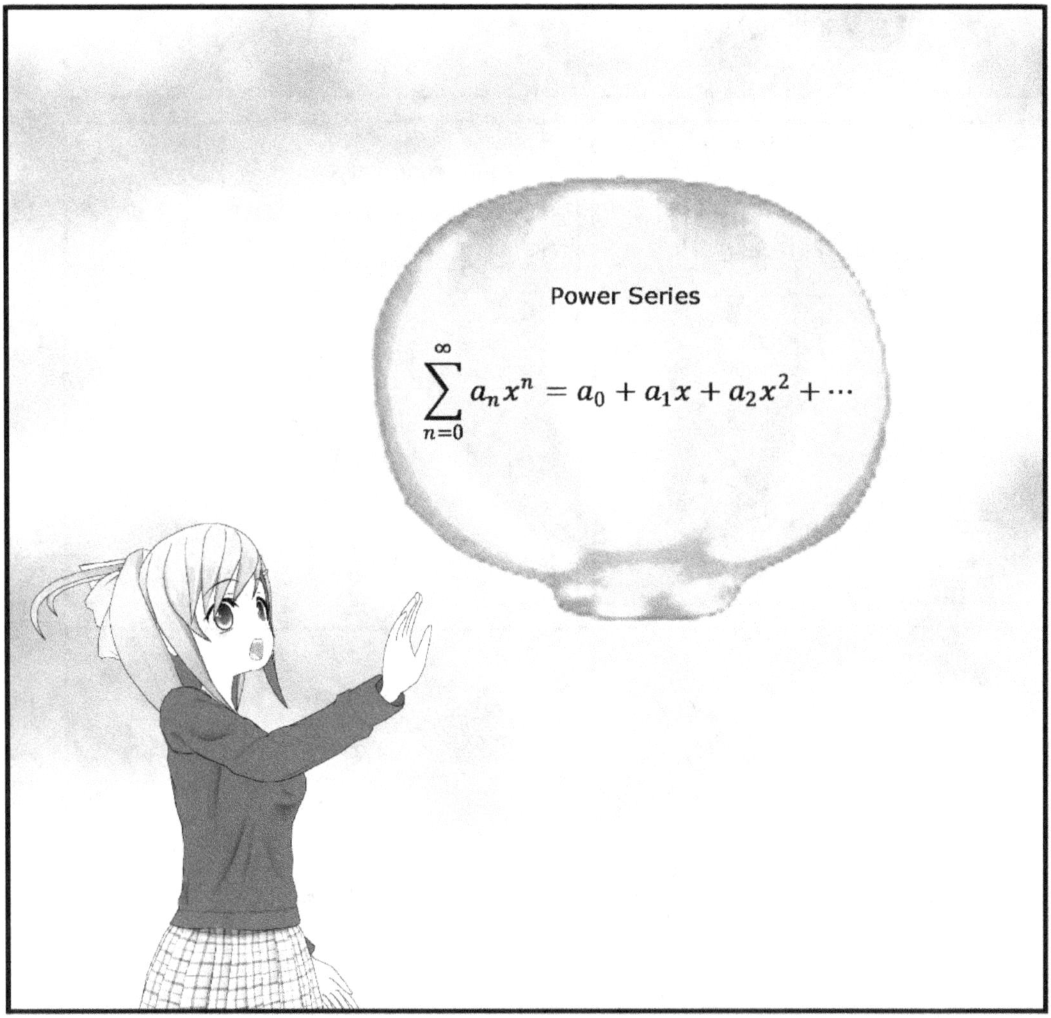

Using the power series method, we suppose that the solution to
$y'' + f(x)y' + g(x)y = r(x)$
is
$$y = \sum_{n=0}^{\infty} a_n x^n$$

Then, we put y,
$$y' = \sum_{n=1}^{\infty} n a_n x^{n-1},$$
$$y'' = \sum_{n=2}^{\infty} n(n-1) a_n x^{n-2}$$
into the original equation.

Then,
we sort the terms,
and find the coefficients.

Clear?

Similar to the method of undetermined coefficients!

Um...
Solve $y'' + xy' - y = 0$.

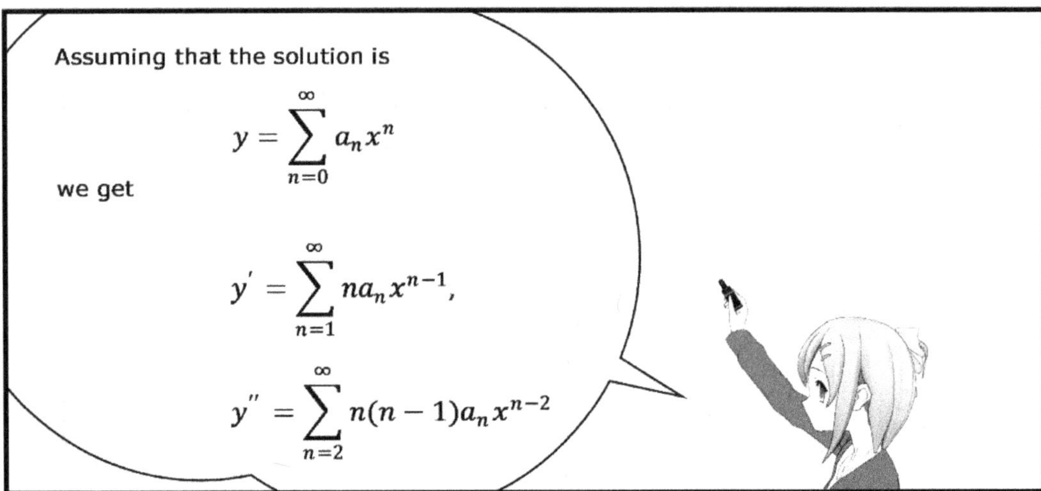

Assuming that the solution is
$$y = \sum_{n=0}^{\infty} a_n x^n$$
we get
$$y' = \sum_{n=1}^{\infty} n a_n x^{n-1},$$
$$y'' = \sum_{n=2}^{\infty} n(n-1) a_n x^{n-2}$$

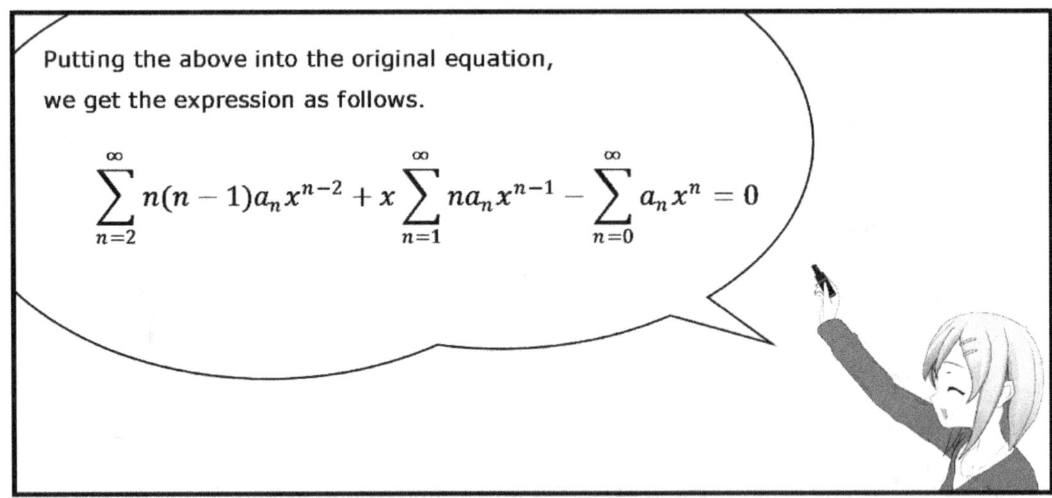

Putting the above into the original equation,
we get the expression as follows.
$$\sum_{n=2}^{\infty} n(n-1) a_n x^{n-2} + x \sum_{n=1}^{\infty} n a_n x^{n-1} - \sum_{n=0}^{\infty} a_n x^n = 0$$

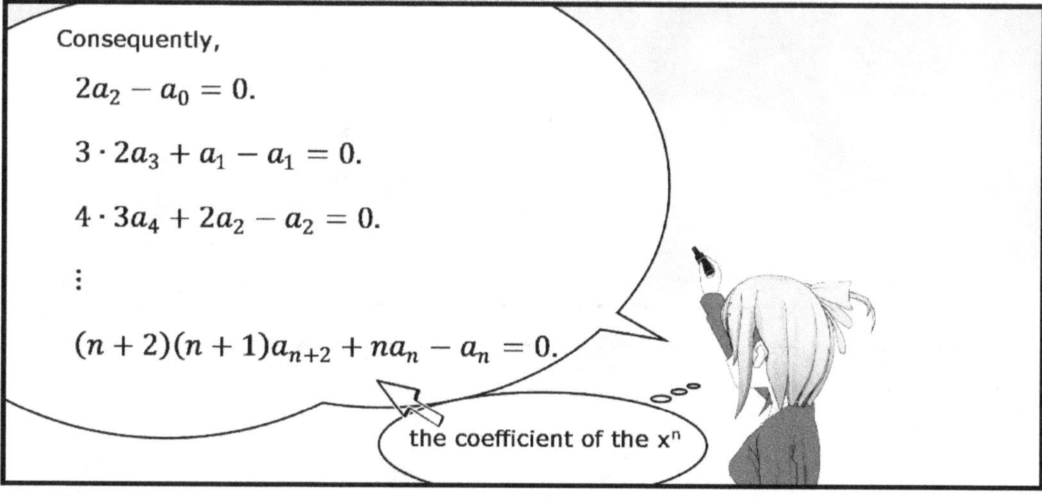

Expanding the Σ, we get

$$2 \cdot 1 a_2 + 3 \cdot 2 a_3 x + 4 \cdot 3 a_4 x^2 + 5 \cdot 4 a_5 x^3 + \cdots$$

$$+ a_1 x + 2 a_2 x^2 + 3 a_3 x^3 + \cdots$$

$$- a_0 - a_1 x - a_2 x^2 - a_3 x^3 - \cdots = 0.$$

Consequently,

$$2 a_2 - a_0 = 0.$$

$$3 \cdot 2 a_3 + a_1 - a_1 = 0.$$

$$4 \cdot 3 a_4 + 2 a_2 - a_2 = 0.$$

$$\vdots$$

$$(n+2)(n+1) a_{n+2} + n a_n - a_n = 0.$$

the coefficient of the x^n

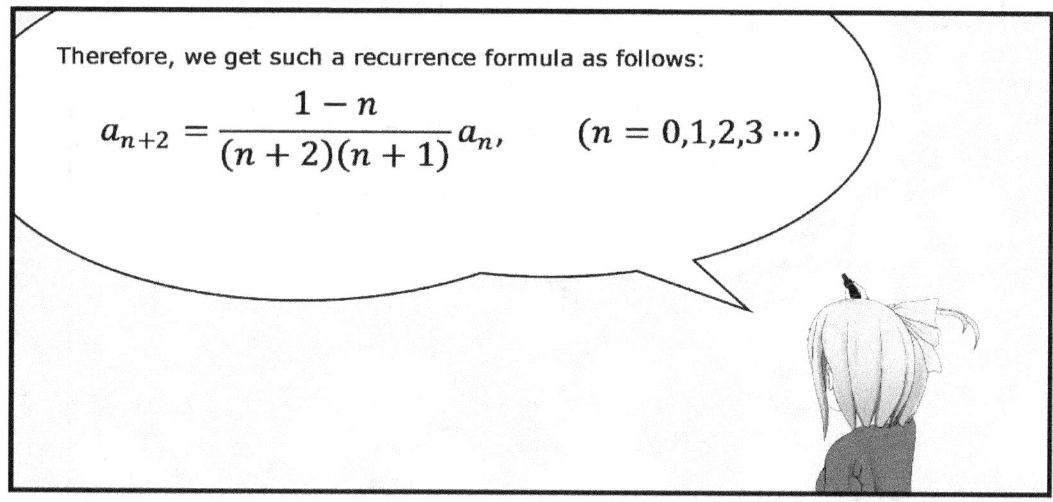

Therefore, we get such a recurrence formula as follows:

$$a_{n+2} = \frac{1-n}{(n+2)(n+1)} a_n, \qquad (n = 0,1,2,3 \cdots)$$

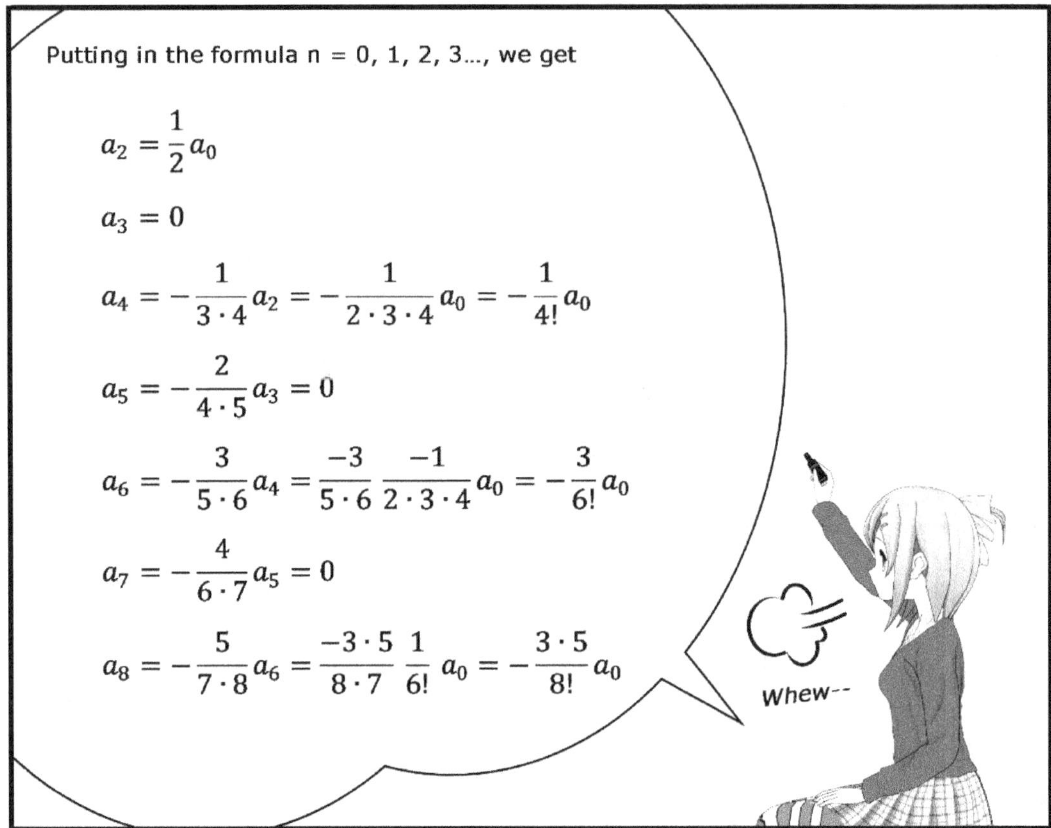

$$a_2 = \frac{1}{2}a_0$$

$$a_3 = 0$$

$$a_4 = -\frac{1}{3\cdot 4}a_2 = -\frac{1}{2\cdot 3\cdot 4}a_0 = -\frac{1}{4!}a_0$$

$$a_5 = -\frac{2}{4\cdot 5}a_3 = 0$$

$$a_6 = -\frac{3}{5\cdot 6}a_4 = \frac{-3}{5\cdot 6}\frac{-1}{2\cdot 3\cdot 4}a_0 = -\frac{3}{6!}a_0$$

$$a_7 = -\frac{4}{6\cdot 7}a_5 = 0$$

$$a_8 = -\frac{5}{7\cdot 8}a_6 = \frac{-3\cdot 5}{8\cdot 7}\frac{1}{6!}a_0 = -\frac{3\cdot 5}{8!}a_0$$

$$y = \sum_{n=0}^{\infty} a_n x^n = a_0 + a_1 x + a_2 x^2 + a_3 x^3 + a_4 x^4 + \cdots$$

$$y = a_0 + a_1 x + \frac{1}{2}a_0 x^2 - \frac{1}{4!}a_0 x^4 + \frac{3}{6!}a_0 x^6 - \frac{3\cdot 5}{8!}a_0 x^8 + \cdots$$

$$= a_1 x + a_0\left(1 + \frac{1}{2}x^2 - \frac{1}{4!}x^4 + \frac{3}{6!}x^6 - \frac{3\cdot 5}{8!}x^8 + \cdots\right)$$

$$y_1(x) \qquad\qquad y_2(x)$$

Now, we get y = a₁y₁(x) + a₀y₂(x),
which is the general solution to the equation given
since y₁(x) and y₂(x) are linearly independent.
Yippee!

Of all the diff-eq
where the coefficients are functions of x,
the most famous are
Legendre's diff-eq and Bessel's diff-eq.

Legendre's diff-eq

$(1 - x^2)y'' - 2xy'$

$+n(n + 1)y = 0$

where n is constant

Bessel's diff-eq

$x^2 y'' + xy'$

$+(x^2 - v^2)y = 0$

where v is
nonnegative real.

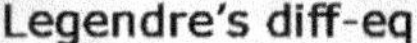

We can find the general solution to this eq
by means of the power series method
as the way you did earlier,
and the solution alone is as follows:

$$y(x) = a_0 y_1(x) + a_1 y_2(x)$$

$$y_1(x) = 1 - \frac{n(n + 1)}{2!} x^2 + \frac{(n - 2)n(n + 1)(n + 3)}{4!} x^4 + \cdots$$

$$y_2(x) = x - \frac{(n - 1)(n + 2)}{3!} x^3 + \frac{(n - 3)(n - 1)(n + 2)(n + 4)}{5!} x^5 + \cdots$$

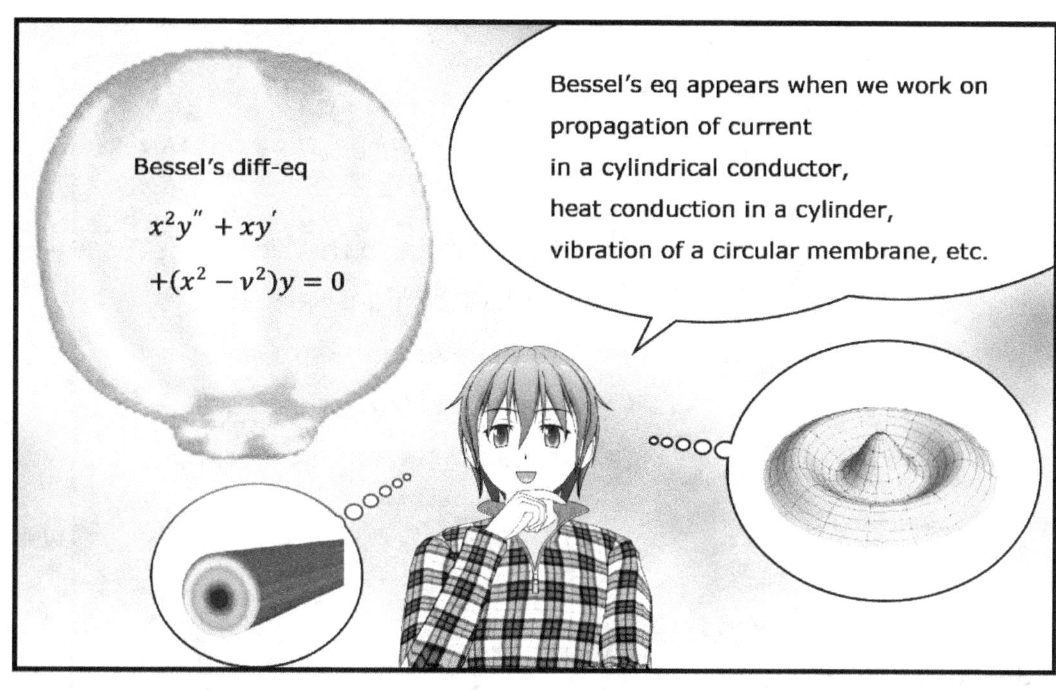

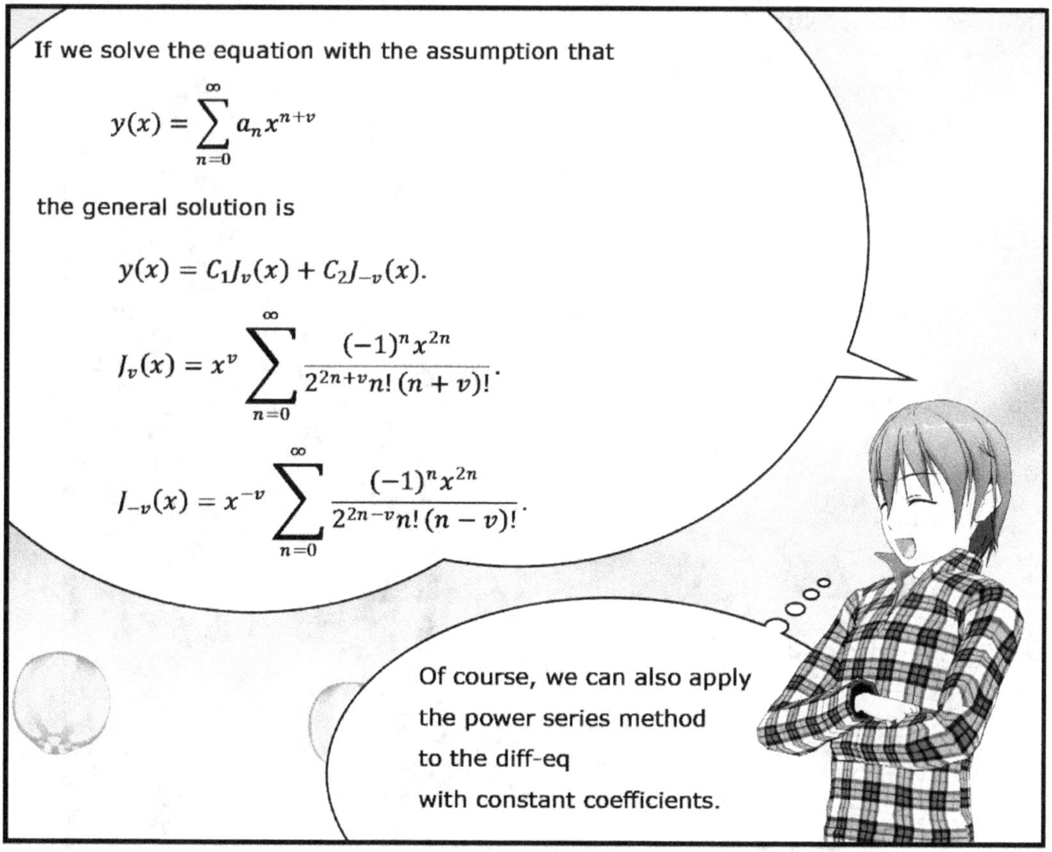

If we solve the equation with the assumption that

$$y(x) = \sum_{n=0}^{\infty} a_n x^{n+v}$$

the general solution is

$$y(x) = C_1 J_v(x) + C_2 J_{-v}(x).$$

$$J_v(x) = x^v \sum_{n=0}^{\infty} \frac{(-1)^n x^{2n}}{2^{2n+v} n! (n+v)!}.$$

$$J_{-v}(x) = x^{-v} \sum_{n=0}^{\infty} \frac{(-1)^n x^{2n}}{2^{2n-v} n! (n-v)!}.$$

Of course, we can also apply the power series method to the diff-eq with constant coefficients.

Give a coin to this youth and let him go. He seems to believe he has to get some materialistic return on doing academic work.
Teacher! What in the world is the good of learning such complicated stuff?
An anonymous youth
Euclid

Quiz 1

Can we solve by the power series method all the 2nd order diff-eq
in the general form of y″ + f(x)y′ + g(x)y = r(x)
where the coefficients are not constant?

Answer

I am afraid not. We can solve particular ones only.
Generally, there are more that are not solvable.
We don't have a general method for solving 2nd order diff-eq
where the coefficients are not constant.
Consequently, we often get the solutions
in approximations via computers.

On the other hand, we do have a general method
for solving 1st order diff-eq in the form of y′ + p(x)y = r(x)
where the coefficients are not constant.
Remember?

$$y(x) = e^{-h}\left[\int e^{h}\cdot r\,dx + C\right]$$

$$\left(h = \int p(x)\,dx\right)$$

All we have to do is just putting the diff-eq into
the above solution formular.

y′ + p(x)y = r(x) is called 1st order linear diff-eq,
and y″ + f(x)y′ + g(x)y = r(x) is called 2nd order linear diff-eq.
These two are the most important in applications of diff-eq.

Laplace Transformations

What is that?

It's a USB flash drive.
Want to try one?
USB

Laplace Transformation

Lapace transformation is a method devised by Laplace
for solving a diff-eq more easily.

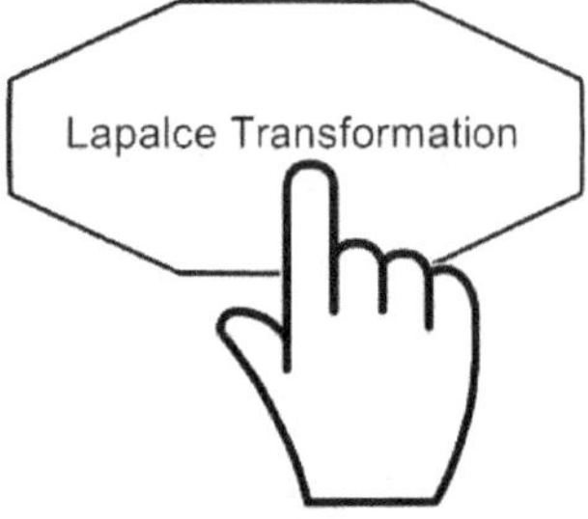

Laplace transformation
on a function f(t) is

$$\mathcal{L}(f) = \int_0^\infty e^{-st} f(t)\,dt$$

The inverse transformation is

$$f(t) = \mathcal{L}^{-1}\mathcal{L}(f)$$

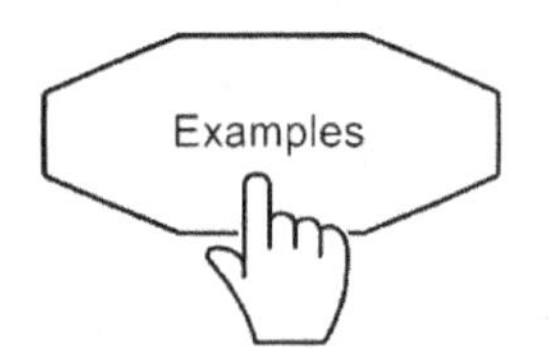

[Example 1]

Find $\mathcal{L}(1)$.

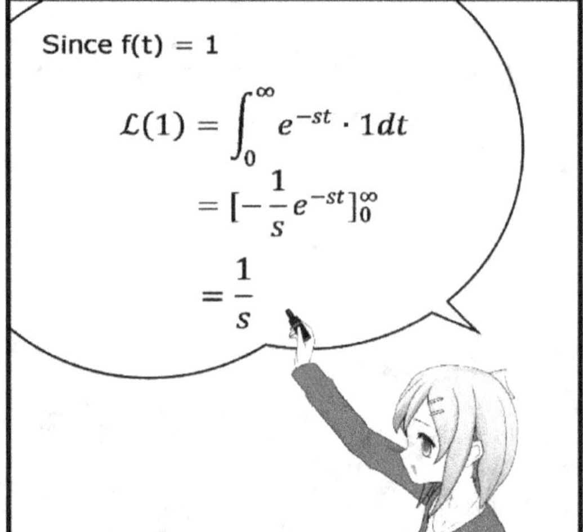

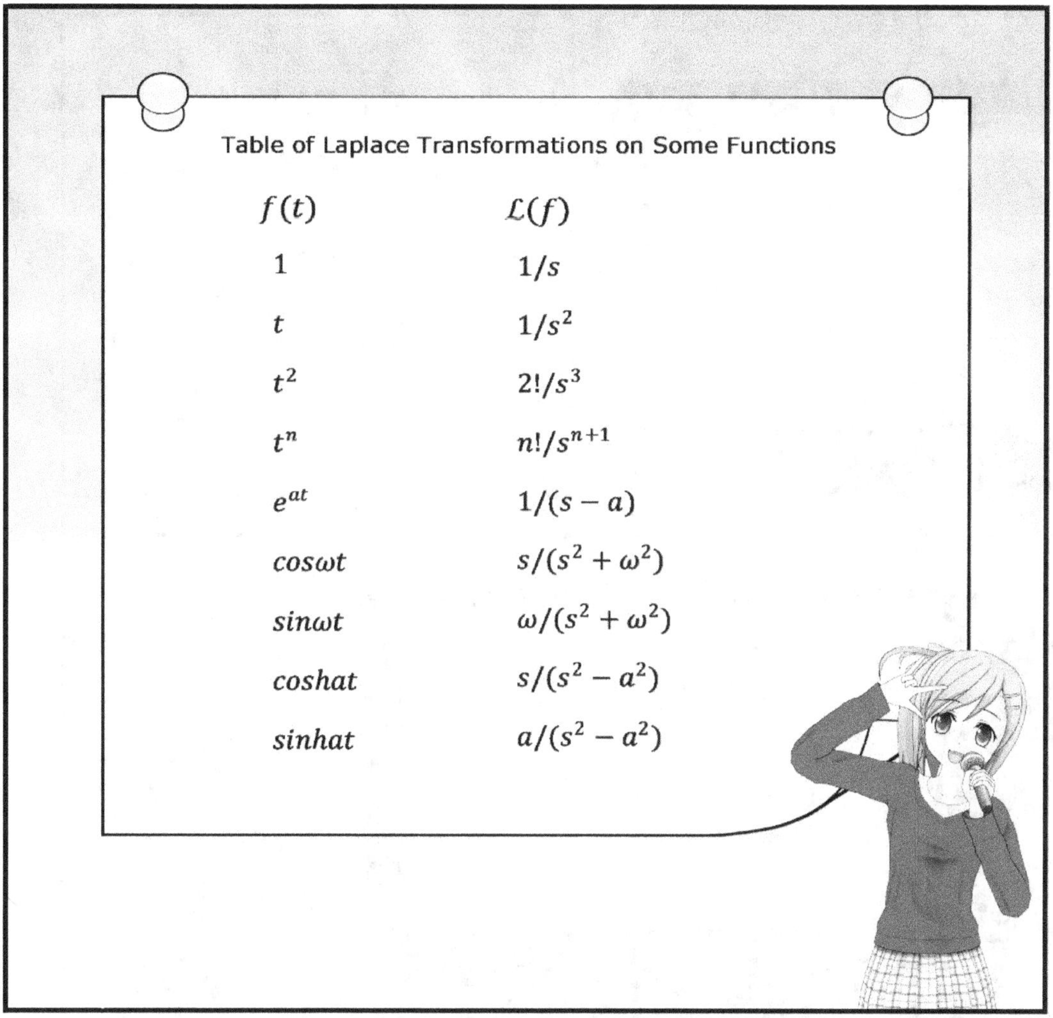

Table of Laplace Transformations on Some Functions

$f(t)$	$\mathcal{L}(f)$
1	$1/s$
t	$1/s^2$
t^2	$2!/s^3$
t^n	$n!/s^{n+1}$
e^{at}	$1/(s-a)$
$cos\omega t$	$s/(s^2+\omega^2)$
$sin\omega t$	$\omega/(s^2+\omega^2)$
$coshat$	$s/(s^2-a^2)$
$sinhat$	$a/(s^2-a^2)$

[Example 2]

Find $\mathcal{L}(f')$.

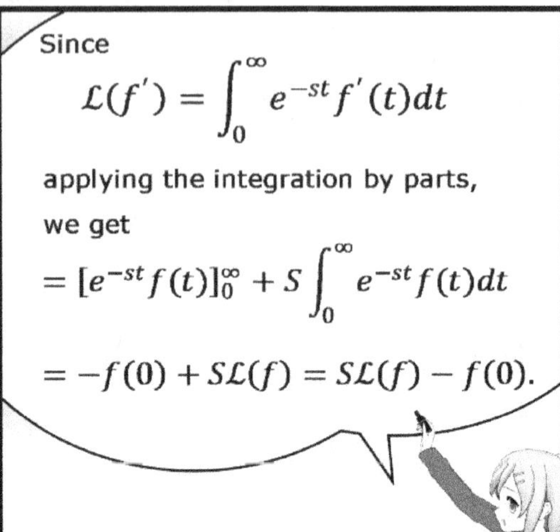
Since
$$\mathcal{L}(f') = \int_0^\infty e^{-st} f'(t)\,dt$$
applying the integration by parts, we get
$$= [e^{-st} f(t)]_0^\infty + S \int_0^\infty e^{-st} f(t)\,dt$$
$$= -f(0) + S\mathcal{L}(f) = S\mathcal{L}(f) - f(0).$$

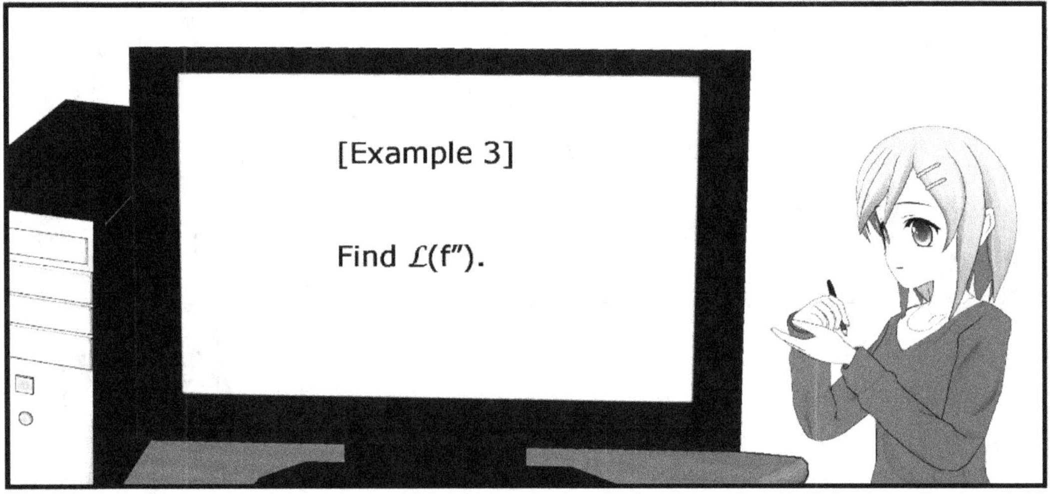
[Example 3]

Find $\mathcal{L}(f'')$.

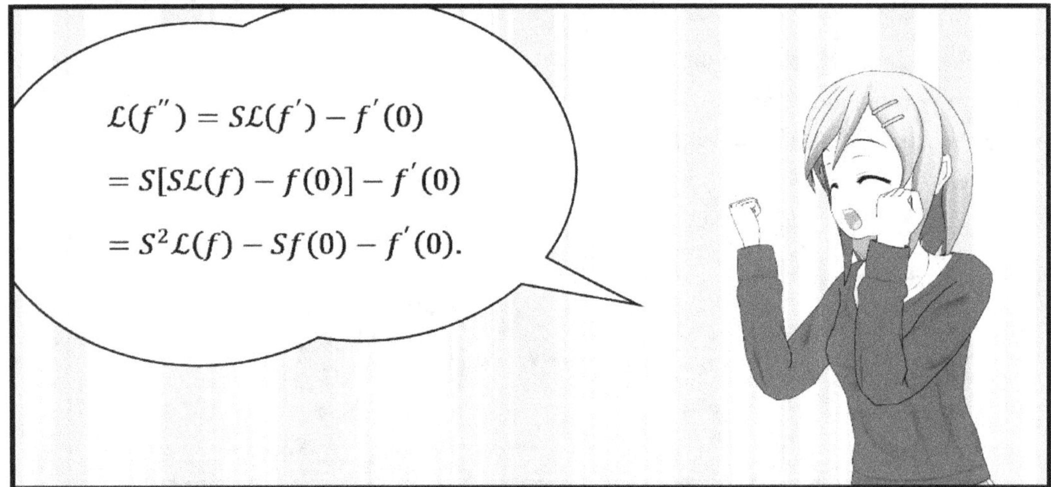
$$\mathcal{L}(f'') = S\mathcal{L}(f') - f'(0)$$
$$= S[S\mathcal{L}(f) - f(0)] - f'(0)$$
$$= S^2\mathcal{L}(f) - Sf(0) - f'(0).$$

[Exercises]

Solve the initial value problem as follows.

$$y'' + 4y' + 3y = 0,$$

$$y(0) = 3, y'(0) = 1.$$

Now, try solving the diff-eq using the Laplace transformation.

Taking Laplace transformation on the equation given, we get

$$\mathcal{L}(y'') + 4\mathcal{L}(y') + 3\mathcal{L}(y) = 0.$$

Now,

$$\mathcal{L}(y'') = S^2\mathcal{L}(y) - Sy(0) - y'(0) = S^2\mathcal{L}(y) - 3S - 1.$$

$$\mathcal{L}(y') = S\mathcal{L}(y) - y(0) = S\mathcal{L}(y) - 3.$$

Putting them back into the equation, we get

$$(S^2\mathcal{L}(y) - 3S - 1) + 4(S\mathcal{L}(y) - 3) + 3\mathcal{L}(y) = 0.$$

$$S^2\mathcal{L}(y) + 4S\mathcal{L}(y) + 3\mathcal{L}(y) = 3S + 13.$$

$$\mathcal{L}(y) = \frac{3S + 13}{s^2 + 4S + 3}.$$

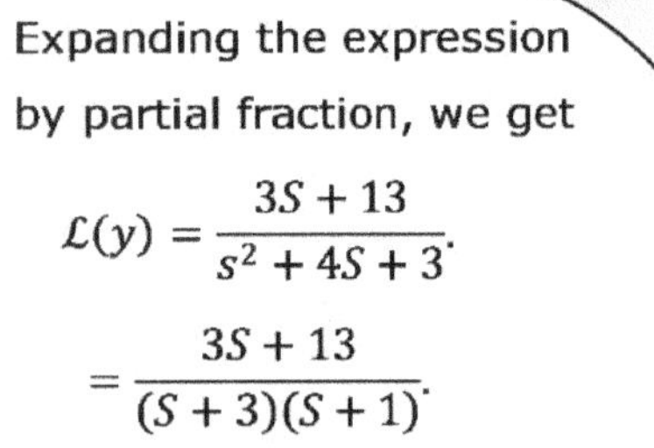

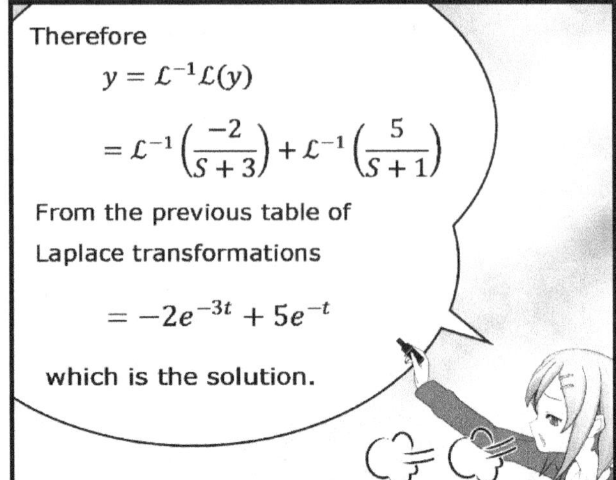

In other words,

if we want to solve a diff-eq

using Laplace transformation:

1) We first take Laplace transformation on the eq given and find the $\mathcal{L}(y)$. Then,

2) We can find the y by taking the inverse transformation on the $\mathcal{L}$, that is, we get $\mathcal{L}^{-1}\mathcal{L}(y) = y$.

Let's go downstairs now.
Yes.

Ting-a-ring ---
1
Higher Order Diff-eq
Are we not going to visit that section?
That floor is for complicated equations that include multi-derivatives.
However, the equations are not used very often for applications.
Let's just get out of Math Pavilion now.

Qulz 1

Find Laplace transformation $\mathcal{L}(f)$ to the function

$$f(t) = e^{\alpha t}, \text{where } s - \alpha > 0.$$

Answer

The definition of Laplace transformation says

$$L(f) = \int_0^\infty e^{-st} f\, dt.$$

By substituting the given function, we get

$$L(f) = \int_0^\infty e^{-st} e^{\alpha t}\, dt$$

$$= \int_0^\infty e^{-(s-\alpha)t}\, dt$$

$$= [-\frac{1}{s-\alpha} e^{-(s-\alpha)t}]_0^\infty$$

$$= -\frac{1}{s-\alpha} e^{-(s-\alpha)\cdot\infty} - \left\{-\frac{1}{s-\alpha} e^{-(s-\alpha)\cdot 0}\right\}$$

$$= 0 - \left(-\frac{1}{s-\alpha}\right)$$

$$= \frac{1}{s-\alpha} \; .$$

Qulz 2

Find the particular solution to y″ − y = 0

using Laplace transformation,

where the initial values are y(0) = 1 and y′(0) = 1.

Answer

By taking Laplace transformation to the given diff eq, we get

$$L(y'') - L(y) = 0. \qquad (*)$$

By substituting

$$L(y'') = s^2 L(y) - sy(0) - y'(0) = s^2 L(y) - s - 1$$

into eq (*), we have

$$s^2 L(y) - s - 1 - L(y) = 0$$
$$(s^2 - 1)L(y) = s + 1$$
$$L(y) = (s+1)/(s^2 - 1)$$
$$= s + 1/((s+1)(s-1))$$
$$= 1/(s-1)$$

By taking inverse Laplace transformation on both sides to find y

$$L^{-1}L(y) = y = L^{-1}\{(1/(s-1)\} = e^t$$

CHAPTER 6
Applications of 2nd Order Diff-eqs
- Mechanical and Electrical Systems -

Applications of 2nd Order Diff-eqs with Constant Coefficients

Undamped
Free Oscillations
An iron ball
is hanging
on a spring.
Try pulling it
and let go of it.
I can see it
oscillating
up and down.
Keep looking at it and
see what'll happen to it.
Finally, it stops moving
after a good while.
If extended or compressed
a little and released,
every object oscillates
in such a manner.
Does this metal spoon, too?

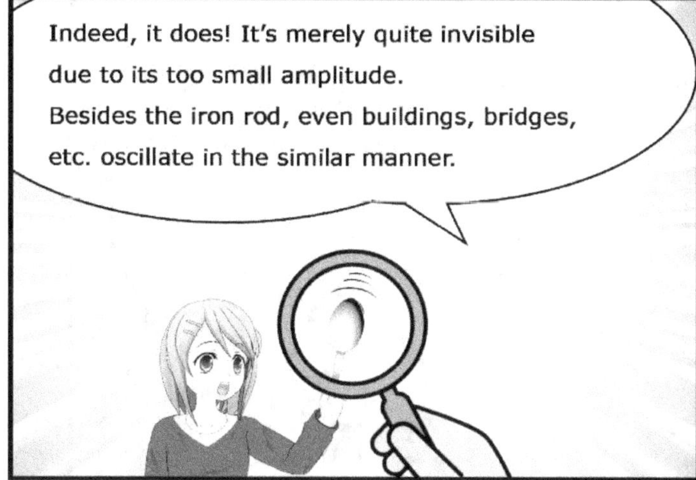
Indeed, it does! It's merely quite invisible
due to its too small amplitude.
Besides the iron rod, even buildings, bridges,
etc. oscillate in the similar manner.

Is that right?
Amazing!

Consequently, such motions of objects
can be simplified in a motion of
an iron ball hanging on a spring.
Such a simplification is called modeling.
Modeling

Simplification for
easy presentation
an expression

The motion of the ball
hanging on a spring
is called a free oscillation.

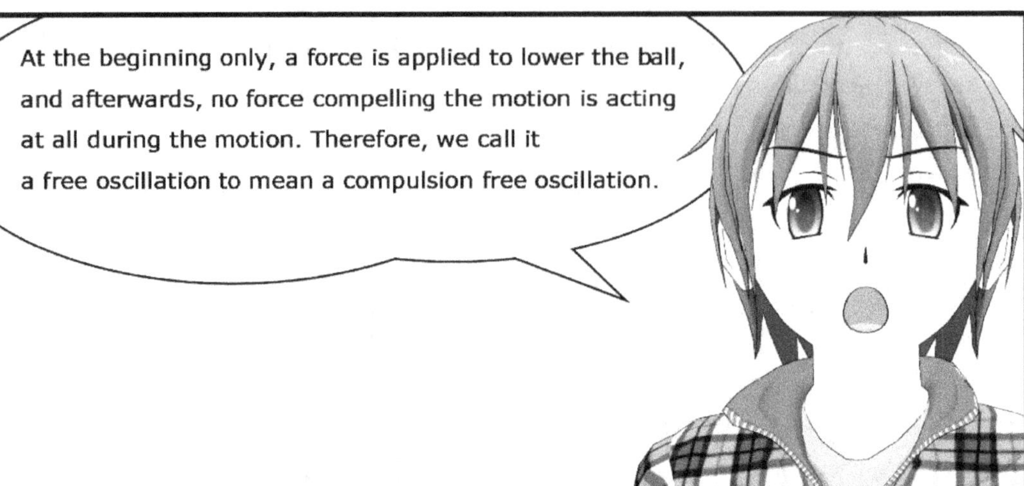
At the beginning only, a force is applied to lower the ball, and afterwards, no force compelling the motion is acting at all during the motion. Therefore, we call it a free oscillation to mean a compulsion free oscillation.

Equation of Motion
Well, what can it be?
What if we put in an expression a motion of such a mechanical system as the ball hanging on a spring?

A motion of any object can be put in F = ma. All objects in the universe obey my equation of motion.
HA HA HA
Gee! Newton, isn't he?

Bye!
Now that you've heard Newton, you can set up the equation of motion, can't you?
Not quite sure about it yet!
I thought I knew about Newton's law, but come to think of it, I don't know much about it!

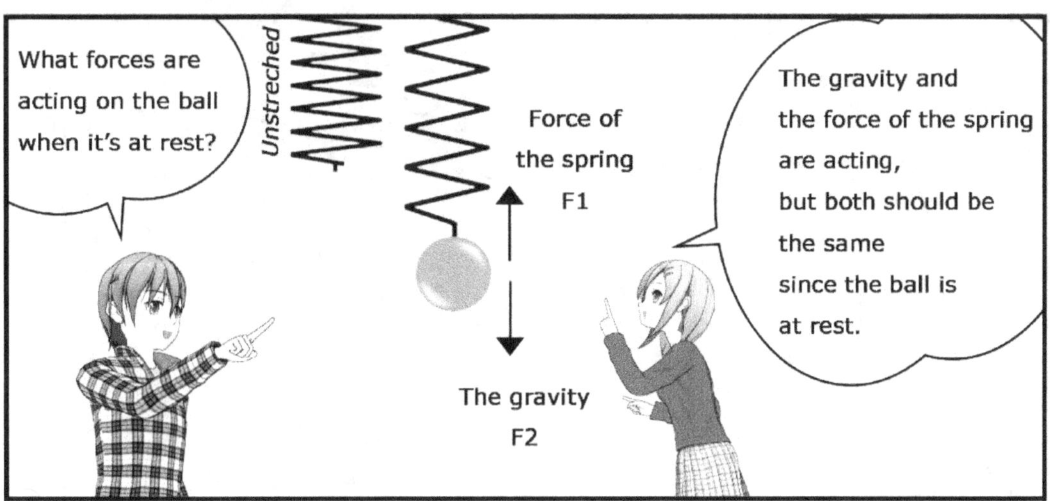

What forces are acting on the ball when it's at rest?
Unstreched
Force of the spring
F1
The gravity
F2
The gravity and the force of the spring are acting, but both should be the same since the ball is at rest.

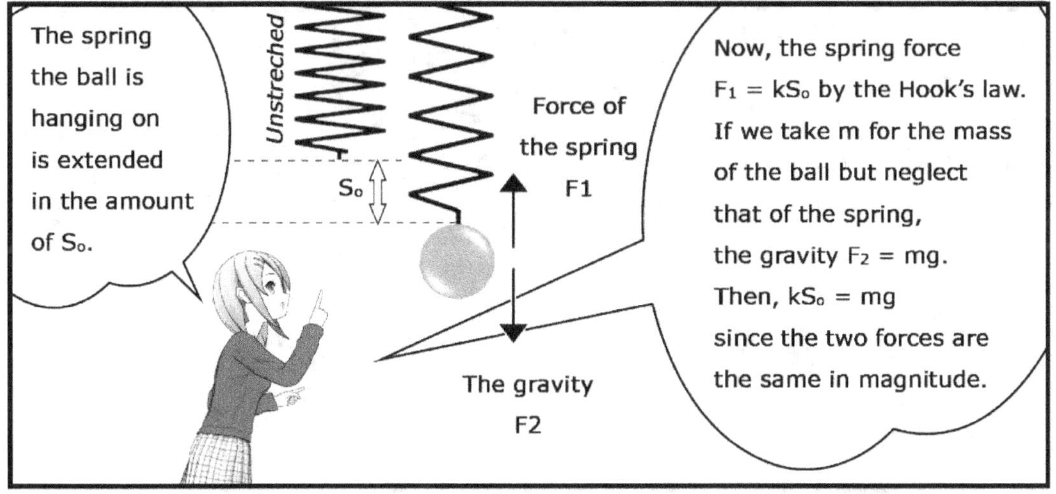

The spring the ball is hanging on is extended in the amount of S_o.
Unstreched
S_o
Force of the spring
F1
The gravity
F2
Now, the spring force $F_1 = kS_o$ by the Hook's law. If we take m for the mass of the ball but neglect that of the spring, the gravity $F_2 = mg$. Then, $kS_o = mg$ since the two forces are the same in magnitude.

Lower the ball
and let go of it.

I can see
it's oscillating.

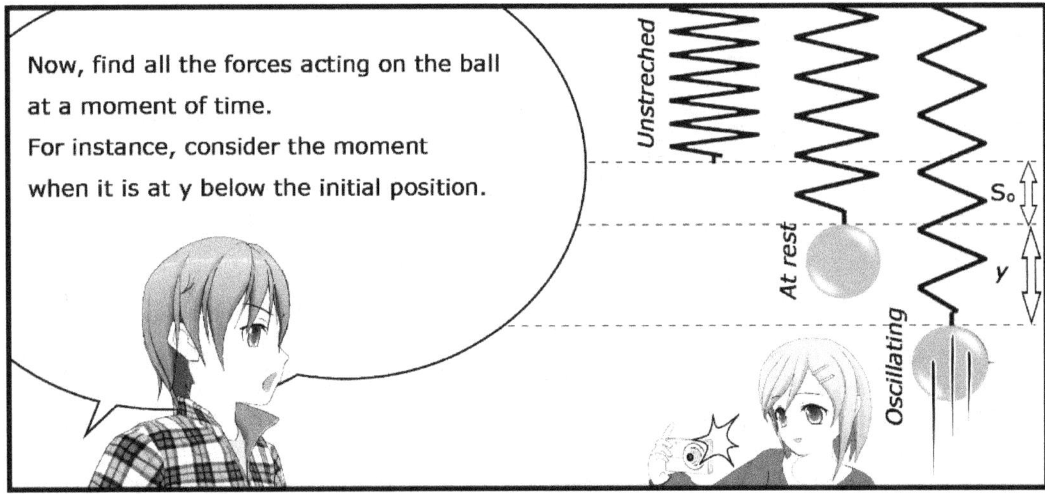
Now, find all the forces acting on the ball
at a moment of time.
For instance, consider the moment
when it is at y below the initial position.
Unstreched
At rest
Oscillating
S_0
y

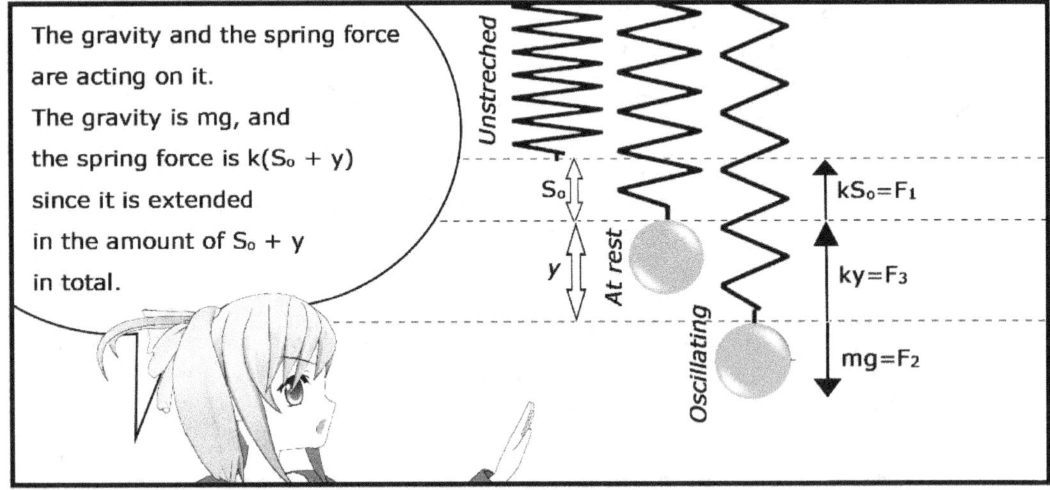
The gravity and the spring force
are acting on it.
The gravity is mg, and
the spring force is $k(S_0 + y)$
since it is extended
in the amount of $S_0 + y$
in total.
Unstreched
At rest
Oscillating
S_0
y
$kS_0 = F_1$
$ky = F_3$
$mg = F_2$

What is the sum
of the forces, then?

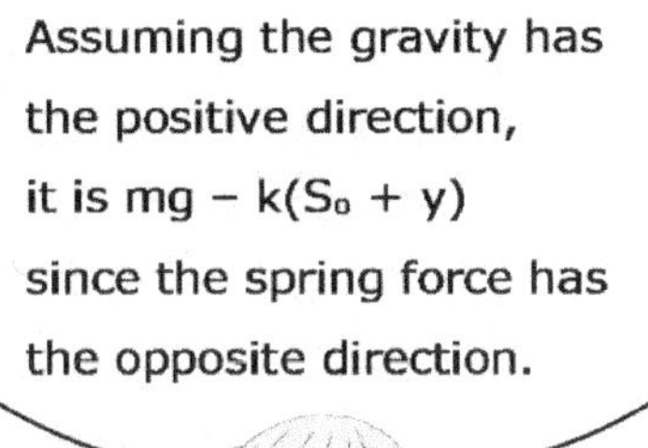

Assuming the gravity has
the positive direction,
it is $mg - k(S_0 + y)$
since the spring force has
the opposite direction.

Now,
since $mg = kS_0$,
it ends up with
$-ky$ only.

That is nothing but the F in F = ma.
That's because the F is the sum of
all the forces acting.

Then, the equation of the motion of
the ball ends up with $-ky = ma$.

It does. Then, since the acceleration a is
the second derivative of the displacement y
with respect to time, that is,

$$a = \frac{d^2y}{dt^2} = y'',$$

we get

$-ky = my''$
$my'' + ky = 0$

where m and k are constant.

Therefore, it gets ended up with
2nd order homogeneous diff-eq.

First of all, we need to put it in the general form of 2nd order homogeneous diff-eq

$$y'' + ay' + by = 0.$$

Dividing both sides of it by m, we get

$$\ddot{y} + \frac{k}{m}y = 0.$$

Assuming the solution is $y = e^{\lambda x}$ and putting it into the above, we get

$$\lambda^2 e^{\lambda x} + \frac{k}{m}e^{\lambda x} = 0.$$

$$(\lambda^2 + \frac{k}{m})e^{\lambda x} = 0.$$

So, the characteristic eq is

$$\lambda^2 + \frac{k}{m} = 0.$$

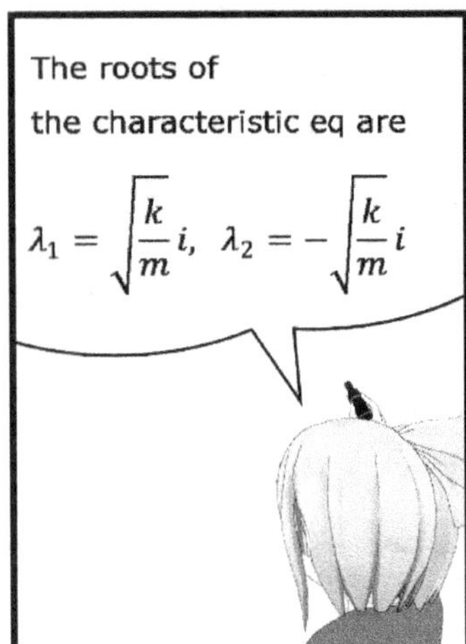

The roots of the characteristic eq are

$$\lambda_1 = \sqrt{\frac{k}{m}}i, \quad \lambda_2 = -\sqrt{\frac{k}{m}}i$$

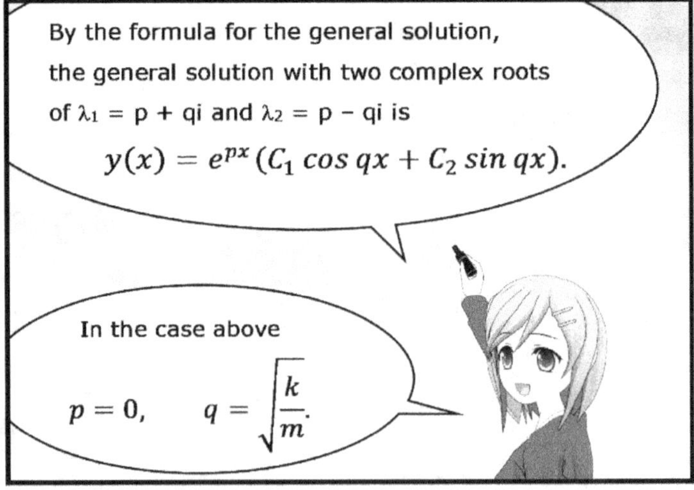

By the formula for the general solution, the general solution with two complex roots of $\lambda_1 = p + qi$ and $\lambda_2 = p - qi$ is

$$y(x) = e^{px}(C_1 \cos qx + C_2 \sin qx).$$

In the case above

$$p = 0, \qquad q = \sqrt{\frac{k}{m}}.$$

Therefore, we can see that the general solution ends up with
$$y(t) = C_1 \cos \sqrt{\frac{k}{m}}\, t + C_2 \sin \sqrt{\frac{k}{m}}\, t.$$

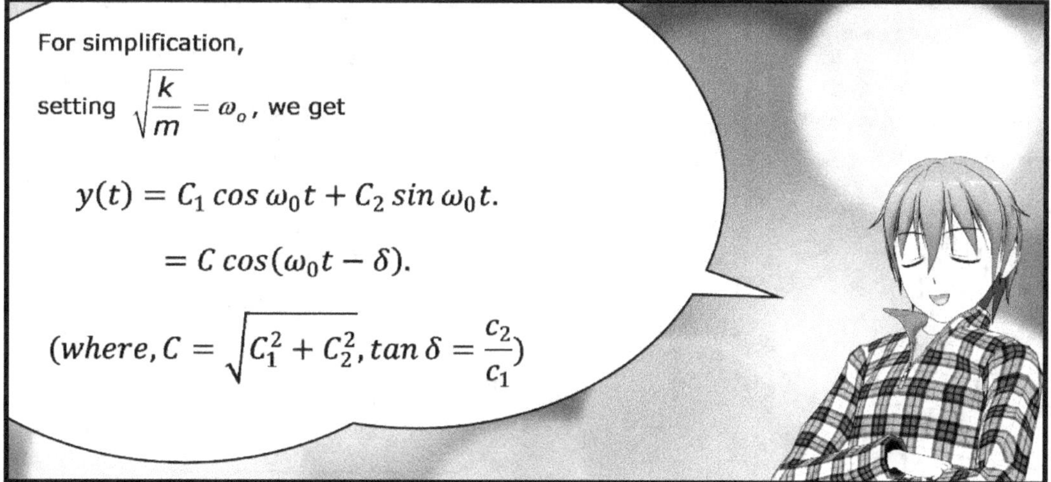

For simplification,
setting $\sqrt{\frac{k}{m}} = \omega_o$, we get
$$y(t) = C_1 \cos \omega_0 t + C_2 \sin \omega_0 t.$$
$$= C \cos(\omega_0 t - \delta).$$
$$\left(where,\ C = \sqrt{C_1^2 + C_2^2},\ \tan \delta = \frac{c_2}{c_1}\right)$$

What is the graph of
y(t) = C cos (ωt - δ)?

y(t)
δ
t

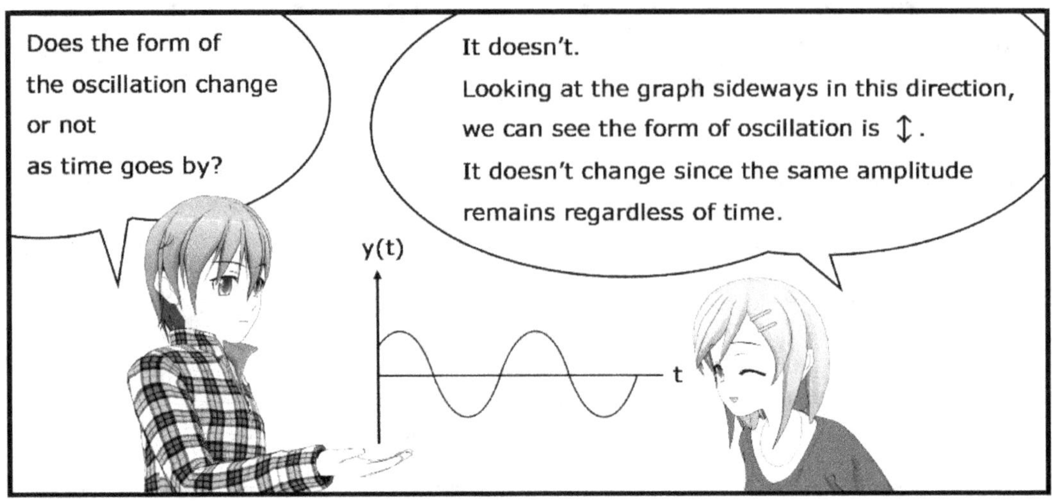

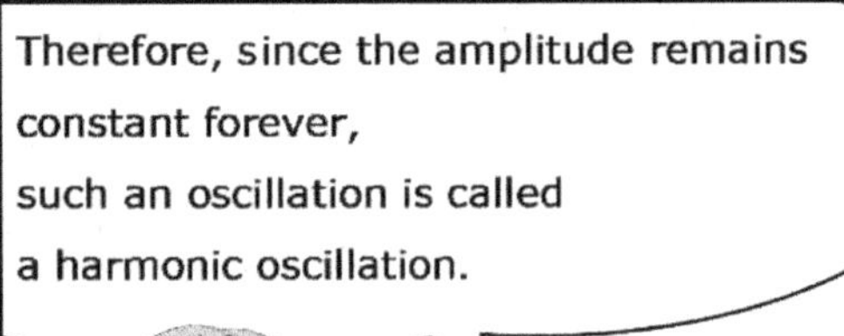

Undamped Free Oscillations

Equation of Motion

$$my'' + ky = 0$$

Characteristic Eq

$$\lambda^2 + \frac{k}{m} = 0$$

General Solution

$$C\cos(\omega_0 t - \delta)$$

y(t)   t

Quiz 1

An iron ball of mass 5.80 kg stretches a spring 8.00 cm.

If we pull down the ball an additional 18.0 cm and let go of it

    a. What will its angular frequency be?

    b. What will its motion be?

Note that the air resistance and the mass of the spring

are to be neglected.

Answer

The motion of the iron ball will be an undamped free oscillation.

a. The angular frequency is given by $\omega_0 = \sqrt{k/m}$.

When the ball is at rest, the gravity mg = the spring force kS_0. So,

$$k = \frac{mg}{S_0} = \frac{(5.80\ kg)(9.80\frac{m}{s^2})}{8.00 \times 10^{-2}m} \approx 711\frac{N}{m}.$$

Therefore,

$$\omega_0 = \sqrt{k/m} = \sqrt{\frac{711\frac{N}{m}}{5.80\ kg}} \approx 11.1\ rad/s$$

Since $\omega_0 = 2\pi f, f = \dfrac{\omega_0}{2\pi f} = \dfrac{11.1}{2\pi} = 1.77$ Hz. So, the iron ball oscillates 1.77 times a second.

b. The general solution of undamped free oscillation is

$$y(t) = C_1 cos\omega_0 t + C_2 sin\omega_0 t. \qquad (*)$$

We have two initial conditions $y(0) = 18.0$ cm and $\dot{y}(0) = 0\ m/s$.

Applying the initial conditions, we get

$$y(0) = 0.180\ m = C_1 cos\omega_0(0) + C_2 sin\omega_0(0) = C_1$$

$$\dot{y}(0) = 0\ m/s = -C_1\omega_0 sin\omega_0(0) + C_2\omega_0 cos\omega_0(0) = C_2\omega_0.\ C_2 = 0$$

Substituting C_1, C_1 and ω_0 into eq $(*)$, we get

$$y(t) = 0.180cos11.1t\ (m).$$

Quiz 2

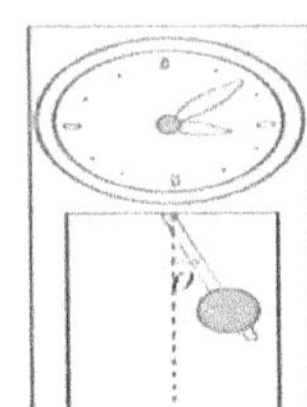

Suppose that we raise a pendulum in a broken clock by an angle θ and let go of it as shown in the figure. Then, put the motion of the pendulum in a diff-eq for θ. Note that the air resistance and the weight of the rod are to be neglected.

Also, assume that θ is small enough to be sin θ ≈ θ.

Answer

If we take m for the mass of the pendulum,

the parallel component of the force of restitution due to the gravity (-mg) is −mg sin θ.

The reason that we put the (-) sign is that the force is for the pendulum's restitution to the standstill.

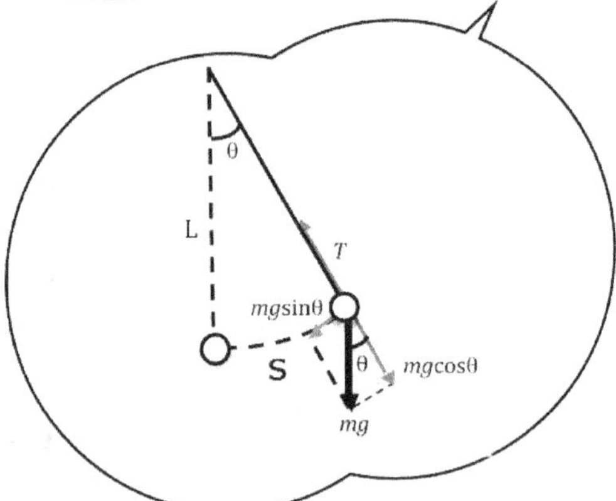

That force is the only force acting on the pendulum.

Taking L for the length of the rod and S for the displacement from the position of standstill, we get $S = L\theta$.

The Newton's second law is F = ma, where

$$F = -mg\sin\theta, \quad a = \frac{d^2S}{dt^2} = \frac{d^2(L\theta)}{dt^2} = \frac{Ld^2\theta}{dt^2}.$$

Therefore

$$-mg\sin\theta = mL\frac{d^2\theta}{dt^2}.$$

So

$$\frac{d^2\theta}{dt^2} + \frac{g}{L}\sin\theta = 0.$$

Since sin θ ≈ θ,

$$\ddot{\theta} + \frac{g}{L}\theta = 0.$$

For reference, the period of the pendulum is $f = \frac{1}{2\pi}\sqrt{\frac{g}{L}}.$

Quiz 3

The equation of motion of the pendulum found in the quiz 2 above, doesn't include m, the mass of the pendulum.
Now, what does that mean?

Answer

- An undamped free oscillation is an ideal oscillation.
- Despite the same undamped free oscillations, in the spring-iron ball model, the force of restitution is due to the spring (-ky), the force does not include the mass m, and after all, the equation of motion, F(= -ky) = my" includes the mass m since the m doesn't get canceled. However, in both cases of the pendulum and the freefall in vacuum, the restitution forces (-mg sin θ and mg) are due to the gravity, the forces include the mass m, the m shows up on both sides of the equation, and after all, the m gets canceled and doesn't get included in the final equation of motion.

Damped Free Oscillations

The typical model offering
a damping force is a dashpot.

To produce a model closer to
the real phenomenon,
we need to include
a dashpot in the model.

Let's make a dashpot.
Fill up about half
the large pot with syrup.

Then, put the small lid
in the pot.

Try pressing
the small lid slow.

This time,
try pressing it fast.
Tee-hee,
tee-hee...

Which of the two is harder,
pressing it slow or fast?
Pressing it fast.

What if we put it
in an expression?

Since the resistance is proportional to
the pressing speed, we can put it in $c\dot{y}$;
however, it will be $-c\dot{y}$ since its direction is
opposite of the direction of motion.
What if we add
the dashpot
to the expression?

Summing up all the acting forces, we get

$$-ky - c\dot{y}.$$

Since the sum is the F in F = ma, we get

$$-ky - c\dot{y} = ma = m\ddot{y}.$$

In other words

$$m\ddot{y} + c\dot{y} + ky = 0.$$

which is the equation of the motion.

That's also a 2nd order homogeneous diff-eq with const-coeffs.

Putting it in the general form, we get

$$\ddot{y} + \frac{c}{m}\dot{y} + \frac{k}{m}y = 0.$$

The characteristic eq is

$$\lambda^2 + \frac{c}{m}\lambda + \frac{k}{m} = 0$$

Therefore, the two roots are

$$\lambda_{1,2} = -\frac{c}{2m} \pm \frac{1}{2m}\sqrt{c^2 - 4mk}.$$

So, as covered in high school, it'll have

i) Two real roots if $c^2 > 4mk$

ii) Two complex roots if $c^2 < 4mk$

iii) A double root if $c^2 = 4mk$

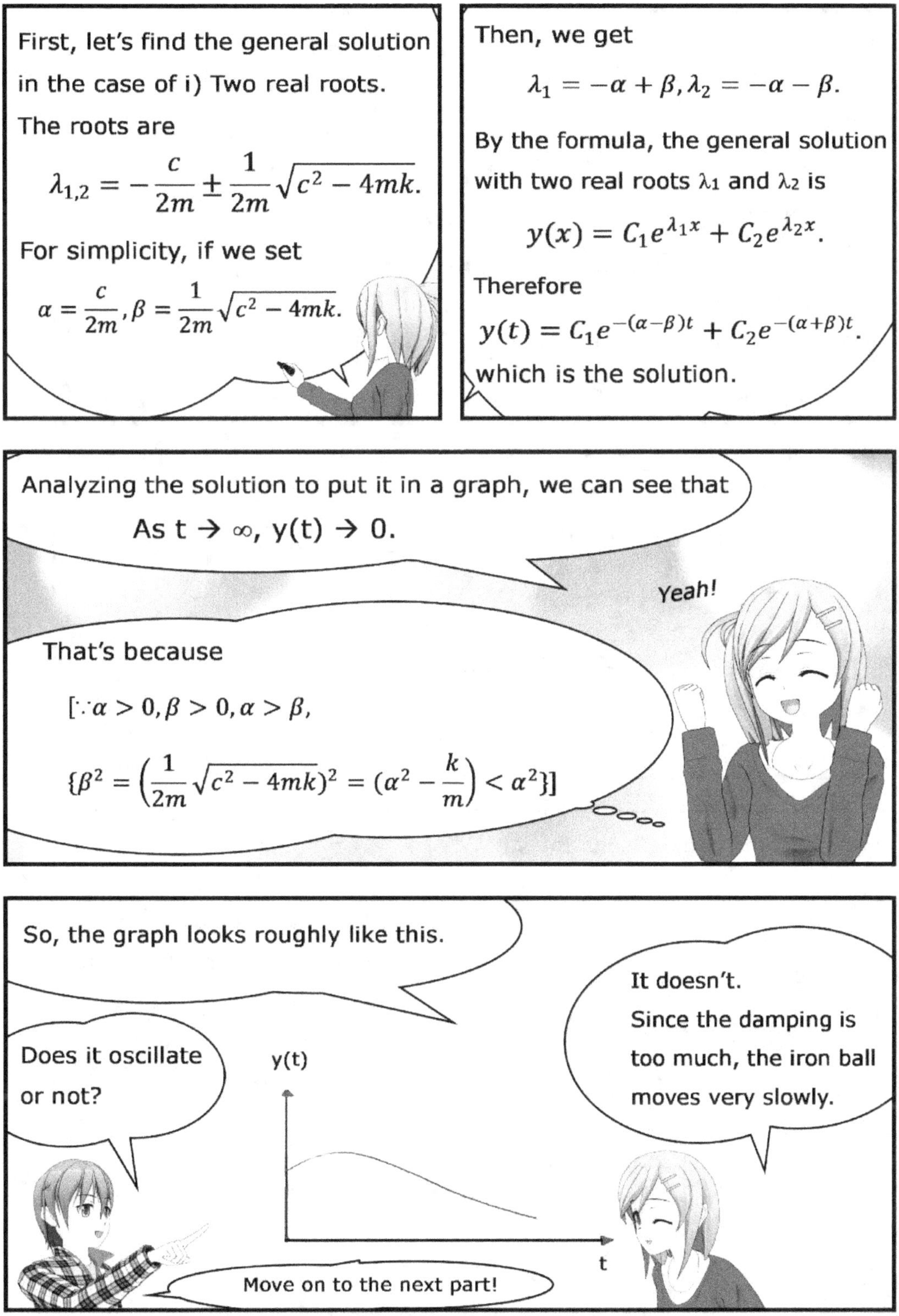

$$\lambda_{1,2} = -\frac{c}{2m} \pm \frac{1}{2m}\sqrt{c^2 - 4mk}.$$

$$\alpha = \frac{c}{2m}, \beta = \frac{1}{2m}\sqrt{c^2 - 4mk}.$$

$$\lambda_1 = -\alpha + \beta, \lambda_2 = -\alpha - \beta.$$

$$y(x) = C_1 e^{\lambda_1 x} + C_2 e^{\lambda_2 x}.$$

$$y(t) = C_1 e^{-(\alpha-\beta)t} + C_2 e^{-(\alpha+\beta)t}.$$

$$[\because \alpha > 0, \beta > 0, \alpha > \beta,$$

$$\{\beta^2 = \left(\frac{1}{2m}\sqrt{c^2 - 4mk}\right)^2 = \left(\alpha^2 - \frac{k}{m}\right) < \alpha^2\}]$$

Secondly, I'll find the general solution in the case of ii) Two complex roots. The roots are

$$\lambda_{1,2} = -\frac{c}{2m} \pm i\,\frac{1}{2m}\sqrt{4mk - c^2}.$$

For simplicity, if we set

$$\alpha = \frac{c}{2m},\ \omega^* = \frac{1}{2m}\sqrt{4mk - c^2},$$

$$\lambda_1 = -\alpha + i\omega^*,$$

$$\lambda_2 = -\alpha - i\omega^*.$$

By the formula, the general solution with two complex roots $\lambda_1 = p + qi$ and $\lambda_2 = p - qi$ is

$$y(x) = e^{px}(C_1 \cos qx + C_2 \sin qx).$$

Therefore, in this case, the general solution is as follows.

$$y(t) = e^{-\alpha t}(C_1 \cos \omega^* t + C_2 \sin \omega^* t).$$

$$= \tilde{c}\,e^{-\alpha t}\cos(\omega^* t - \delta).$$

$$\left(\text{where}, C = \sqrt{C_1^2 + C_2^2},\ \tan \delta = \frac{C_2}{C_1}\right)$$

The graph of the general solution is

Does it oscillate or not?

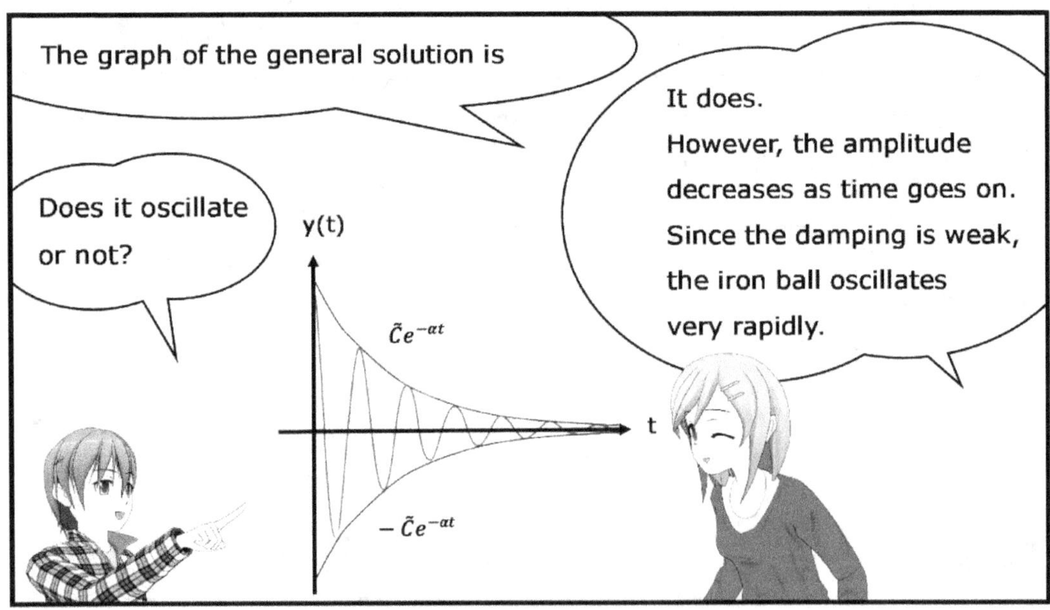

It does.
However, the amplitude decreases as time goes on. Since the damping is weak, the iron ball oscillates very rapidly.

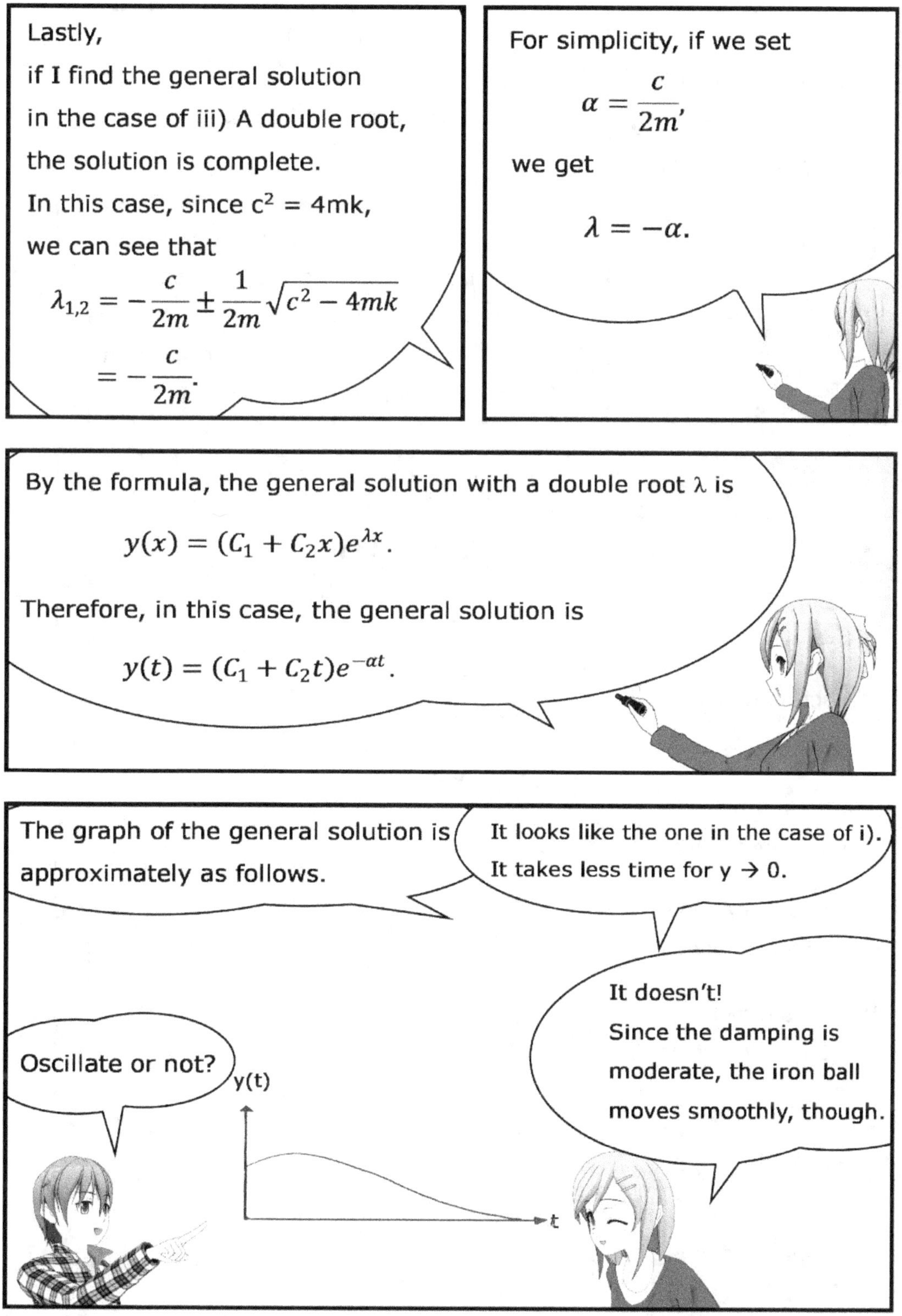

$$\lambda_{1,2} = -\frac{c}{2m} \pm \frac{1}{2m}\sqrt{c^2 - 4mk}$$

$$= -\frac{c}{2m}.$$

$$\alpha = \frac{c}{2m},$$

$$\lambda = -\alpha.$$

$$y(x) = (C_1 + C_2 x)e^{\lambda x}.$$

$$y(t) = (C_1 + C_2 t)e^{-\alpha t}.$$

We call:
The case i) over-damping
The case ii) under-damping
The case iii) critical damping
Now, how do we find the constants C_1 and C_2 included in the general solutions?
We find them by means of the initial conditions. We are normally given two kinds of initial conditions, that is, the initial displacement and the initial velocity.

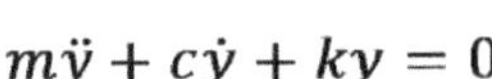

Tantara tantanta tantara

Damped Free Oscillations

Equation of Motion

$m\ddot{y} + c\dot{y} + ky = 0$

Characteristic Eq.

$\lambda^2 + \dfrac{c}{m}\lambda + \dfrac{k}{m} = 0$

i) Two real roots
= Over-damping

ii) Two complex roots
= Under-damping

iii) A double root
= Critical damping

General Solution

$C_1 e^{-(\alpha-\beta)t} + C_2 e^{-(\alpha+\beta)t}$

$\tilde{C} e^{-\alpha t} \cos(\omega^* t - \delta)$

$(C_1 + C_2 t)e^{-\alpha t}$

Graph

Congratulations on your passing the Free Oscillations section!

The most incomprehensible thing about the world is that it is comprehensible.
Albert Einstein

Quiz 1

An iron ball of mass 2.5 kg hangs on a spring of modulus 10 N/m and the iron ball is connected to a dashpot.

The initial displacement is 0.5 m and the initial velocity is 0 m/s.

Find the general solution expressing the motion of the iron ball if the system has damping given by

a. $c = 20$ kg/sec
b. $c = 10$ kg/sec
c. $c = 5$ kg/sec.

Answer

The motion of the iron ball is undamped free oscillation.

The eq. of motion is given by $m\ddot{y} + c\dot{y} + ky = 0$.

a. Substituting the given values to the eq. of motion, we get

$$2.5\ddot{y} + 20\dot{y} + 10y = 0.$$

Dividing both sides by 2.5, we have

$$\ddot{y} + 8\dot{y} + 4y = 0.$$

So, the characteristic eq. is

$$\lambda^2 + 8\lambda + 4 = 0$$

Two real roots are

$$\lambda = \frac{-8 \pm \sqrt{64 - 16}}{-2} \approx \frac{-8 \pm 7}{2} = \begin{cases} -0.5 \\ -7.5 \end{cases}.$$

Therefore, the general solution $y(t)$ is

$$y(t) = C_1 e^{-0.5t} + C_2 e^{-7.5t}.$$

Applying initial conditions, $y(0) = 0.5$ and $\dot{y}(0) = 0$

$$y(0) = C_1 e^{-0.5(0)} + C_2 e^{-7.5(0)} = C_1 + C_2 = 0.5$$

$$\dot{y}(t) = -0.5C_1 e^{-0.5t} - 7.5C_2 e^{-7.5t}$$

$$\dot{y}(0) = -0.5C_1 e^{-0.5(0)} - 7.5C_2 e^{-7.5(0)}$$

$$= -0.5C_1 - 7.5C_2 = 0$$

we get $C_1 = 0.54$ and $C_2 = -0.04$.

Substituting these to the general solution $y(t)$, we get

$$y(t) = 0.54e^{-0.5t} - 0.04e^{-7.5t} \ (m).$$

Answer

b. Substituting the given values to the eq. of motion, we get

$$2.5\ddot{y} + 10\dot{y} + 10y = 0.$$

Dividing both sides by 2.5, we have

$$\ddot{y} + 4\dot{y} + 4y = 0.$$

So, the characteristic eq. is

$$\lambda^2 + 4\lambda + 4 = 0.$$

A double root is $\lambda = -2$.

Therefore, the general solution $y(t)$ is

$$y(t) = (C_1 + C_2 t)e^{-2t}.$$

Applying initial conditions, $y(0) = 0.5$ and $\dot{y}(0) = 0$

$$y(0) = (C_1 + C_2(0))e^{-2(0)} = C_1 = 0.5$$

$$\dot{y}(t) = C_2 e^{-2t} - 2(C_1 + C_2 t)e^{-2t}$$

$$\dot{y}(0) = C_2 e^{-2(0)} - 2(C_1 + C_2(0))e^{-2(0)}$$

$$= C_2 - 2C_1 = 0$$

we get $C_1 = 0.5$ and $C_2 = 1$.

Substituting these to the general solution $y(t)$, we get

$$y(t) = (0.5 + t)e^{-2t} \ (m).$$

c. Substituting the given values to the eq. of motion, we get

$$2.5\ddot{y} + 5\dot{y} + 10y = 0.$$

Dividing both sides by 2.5, we have

$$\ddot{y} + 2\dot{y} + 4y = 0.$$

So, the characteristic eq. is

$$\lambda^2 + 2\lambda + 4 = 0.$$

Two complex roots are $\lambda_1 = -1 + 1.8\,i$ and $\lambda_2 = -1 - 1.8\,i$.

Therefore, the general solution $y(t)$ is

$$y(t) = e^{-t}(C_1 \cos 1.8t + C_2 \sin 1.8t).$$

Applying initial conditions, $y(0) = 0.5$ and $\dot{y}(0) = 0$

we get $C_1 = 0.5$ and $C_2 = 0.28$.

Substituting these to the general solution $y(t)$, we get

$$y(t) = e^{-t}(0.5\cos 1.8t + 0.28\sin 1.8t) \ (m).$$

Quiz 2

The speedometer of a car is designed to be in the state of critical damping in the damped free oscillation.
What is the reason?

Answer

Let's consider the case where we reduce the speed from 60 Km/hr to 40 Km/hr.

Over-damping	Critical damping	Under-damping

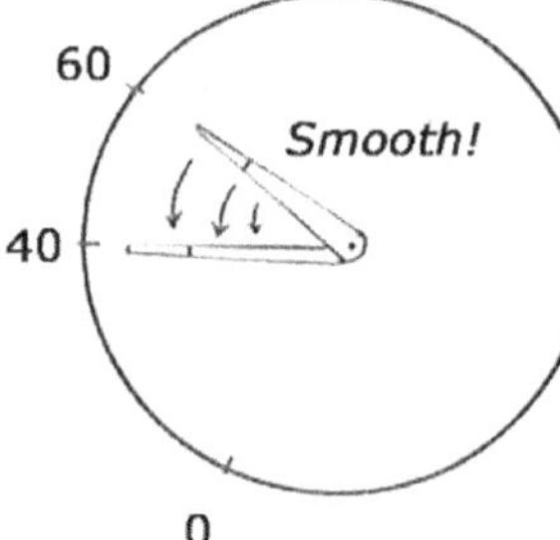

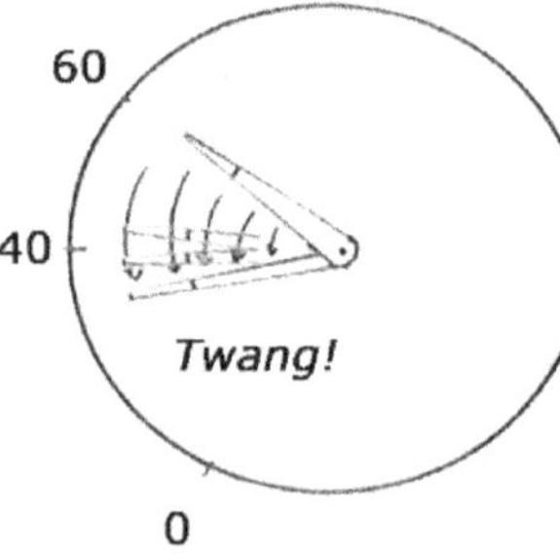

The resistance to the movement of the index is too much, and it takes too long for the index of the speedometer to move from 60 to 40.

The resistance is moderate, and the index moves smoothly from 60 to 40.

The resistance is insufficient, and the index vibrates.

For instance, if we imagine that each speedometer is filled with hot pepper past, soy sauce, or air, we can easily get the idea.

Besides the speedometer in a car, most meters are designed to be in the state of critical damping.

Quiz 3

In the previous quiz on the motion of
the pendulum of a broken clock,
assume that the damping force
due to the air resistance is
given in $cL\dot{\theta}$ (c is the damping constant),
and put the equation of the motion
of the pendulum in a diff-eq for θ.

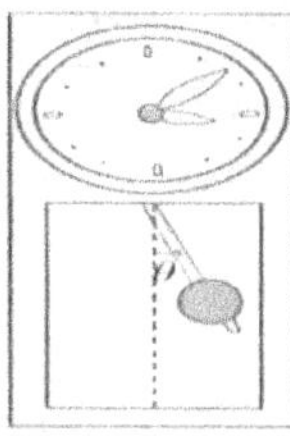

Answer

In the Newton's second law F = ma,
the F is the sum of all the acting forces:

$$-cL\dot{\theta} - mg\sin\theta.$$

For ma:

$$m\frac{d^2S}{dt^2} = m\frac{d^2(L\theta)}{dt^2} = mL\frac{d^2\theta}{dt^2} = mL\ddot{\theta}.$$

Therefore

$$mL\ddot{\theta} = -cL\dot{\theta} - mg\sin\theta.$$

$$\ddot{\theta} + \frac{c}{m}\dot{\theta} + \frac{g}{L}\sin\theta = 0.$$

If $\sin 0 \approx 0$, then

$$\ddot{\theta} + \frac{c}{m}\dot{\theta} + \frac{g}{L}\theta = 0.$$

Applications of 2nd Order Nonhomogeneous Diff-eqs with Const-coeffs I

The oscillations in this kind are called forced oscillations. That's because you are forcing the iron ball to oscillate by applying forces upward and downward.

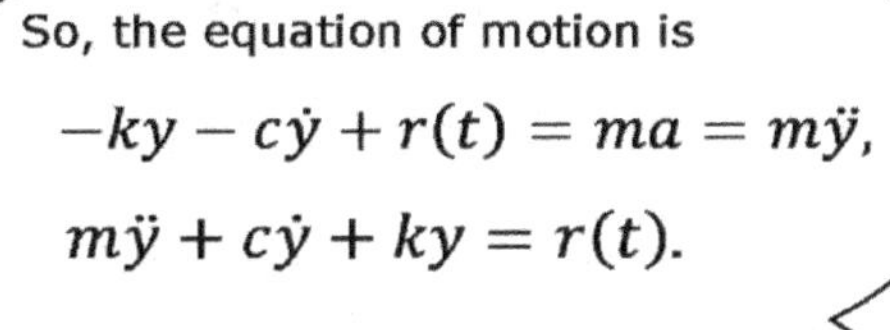

$$-ky - c\dot{y} + r(t) = ma = m\ddot{y},$$
$$m\ddot{y} + c\dot{y} + ky = r(t).$$

If the external force r(x) is either a strain or a propulsion only, the iron ball won't oscillate and it will be flat. So, assume the external force is $F_0 \cos \omega t$ which changes the direction periodically, and solve $m\ddot{y} + c\dot{y} + ky = F_0 cos\omega t.$

The general solution to nonhomogeneous diff-eq, y(t) is
$$y(t) = y_h(t) + y_p(t),$$
where y_h(t) is the general solution to homogeneous diff-eq and y_p(t) is the particular solution to nonhomogeneous diff-eq.

By the way, y_h(t), that is,
$$my'' + cy' + ky = 0$$
the general solution to the above is already known to us.

So, we need to find y_p(t) only. Normally, we find y_p(t) by the undetermined coefficient method. Since we have $r(t) = F_0 cos\omega t,$

from the table found on the 3rd floor in Math Pavilion, let's assume that
$$y_p(t) = acos\omega t + bsin\omega t.$$

Then, put y_p into
$$m\ddot{y} + c\dot{y} + ky = F_0\cos\omega t$$
and find the undetermined coefficients, a and b.

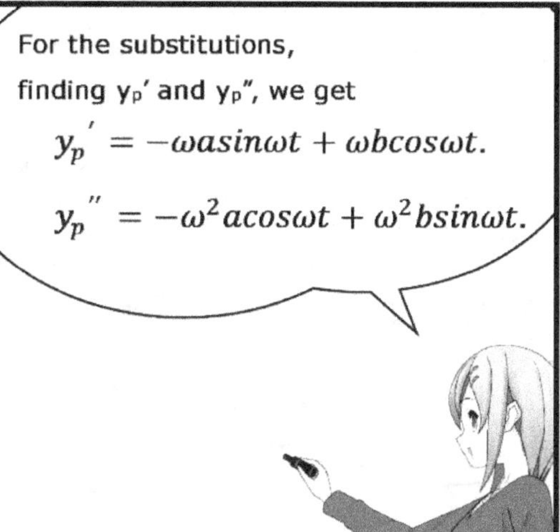

For the substitutions, finding $y_p{}'$ and $y_p{}''$, we get
$$y_p{}' = -\omega a\sin\omega t + \omega b\cos\omega t.$$
$$y_p{}'' = -\omega^2 a\cos\omega t + \omega^2 b\sin\omega t.$$

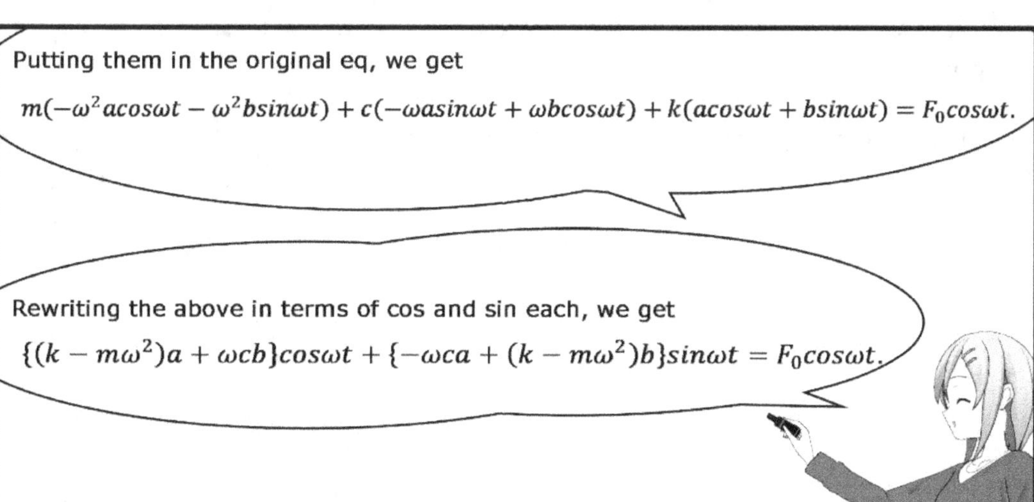

Putting them in the original eq, we get
$$m(-\omega^2 a\cos\omega t - \omega^2 b\sin\omega t) + c(-\omega a\sin\omega t + \omega b\cos\omega t) + k(a\cos\omega t + b\sin\omega t) = F_0\cos\omega t.$$
Rewriting the above in terms of cos and sin each, we get
$$\{(k - m\omega^2)a + \omega cb\}\cos\omega t + \{-\omega ca + (k - m\omega^2)b\}\sin\omega t = F_0\cos\omega t.$$

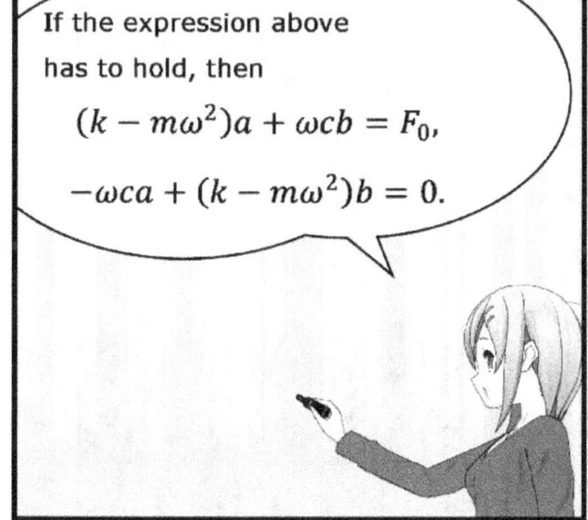

If the expression above has to hold, then
$$(k - m\omega^2)a + \omega cb = F_0,$$
$$-\omega ca + (k - m\omega^2)b = 0.$$

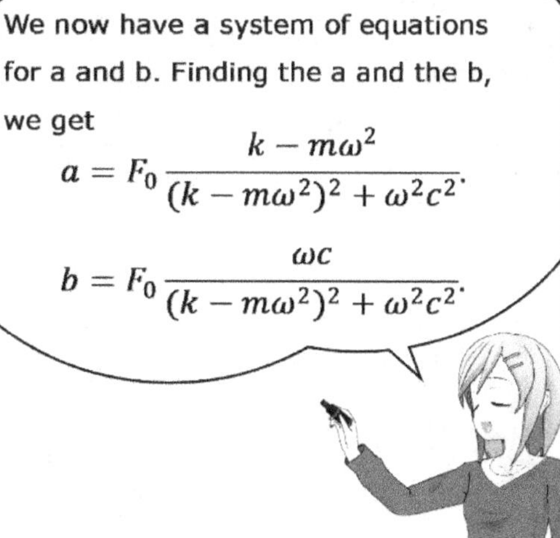

We now have a system of equations for a and b. Finding the a and the b, we get
$$a = F_0\frac{k - m\omega^2}{(k - m\omega^2)^2 + \omega^2 c^2}.$$
$$b = F_0\frac{\omega c}{(k - m\omega^2)^2 + \omega^2 c^2}.$$

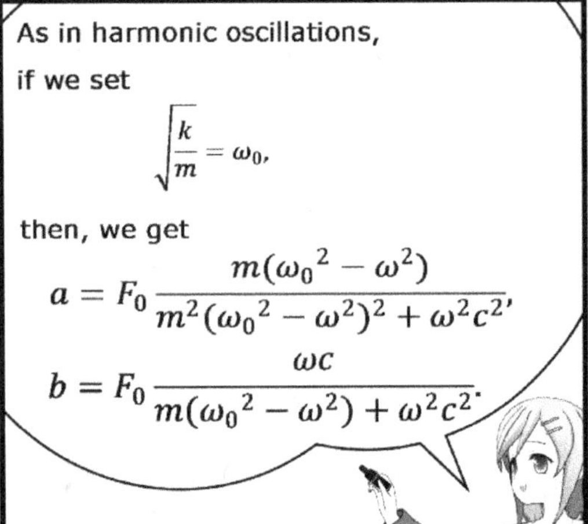

As in harmonic oscillations, if we set
$$\sqrt{\frac{k}{m}} = \omega_0,$$
then, we get
$$a = F_0 \frac{m(\omega_0{}^2 - \omega^2)}{m^2(\omega_0{}^2 - \omega^2)^2 + \omega^2 c^2},$$
$$b = F_0 \frac{\omega c}{m(\omega_0{}^2 - \omega^2) + \omega^2 c^2}.$$

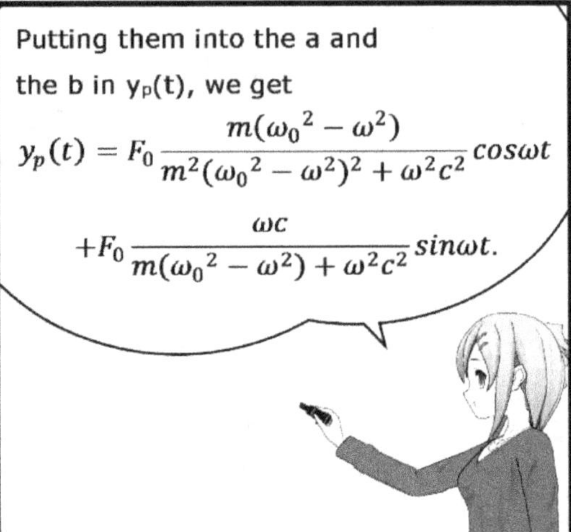

Putting them into the a and the b in $y_p(t)$, we get
$$y_p(t) = F_0 \frac{m(\omega_0{}^2 - \omega^2)}{m^2(\omega_0{}^2 - \omega^2)^2 + \omega^2 c^2} cos\omega t$$
$$+F_0 \frac{\omega c}{m(\omega_0{}^2 - \omega^2) + \omega^2 c^2} sin\omega t.$$

So, the general solution is
$$y(t) = y_h(t) + F_0 \frac{m(\omega_0{}^2 - \omega^2)}{m^2(\omega_0{}^2 - \omega^2)^2 + \omega^2 c^2} cos\omega t + F_0 \frac{\omega c}{m(\omega_0{}^2 - \omega^2) + \omega^2 c^2} sin\omega t.$$

Great job!
Let's now consider the general solution in two separate aspects of being damped and undamped.
Undamped Forced Oscillations

Undamped Forced Oscillations
Cut the dashpot away!
It's already done.
Let's begin with the aspect of being undamped.

What is the eq of motion in undamped forced oscillations?
Since the damping constant c = 0, we get
$m\ddot{y} + ky = F_0 cos\omega t.$

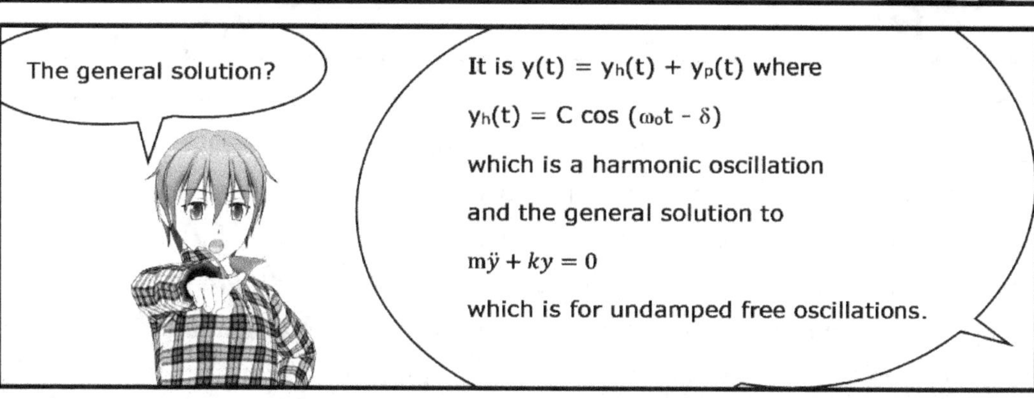

The general solution?
It is y(t) = yh(t) + yp(t) where
yh(t) = C cos (ω0t - δ)
which is a harmonic oscillation
and the general solution to
$m\ddot{y} + ky = 0$
which is for undamped free oscillations.

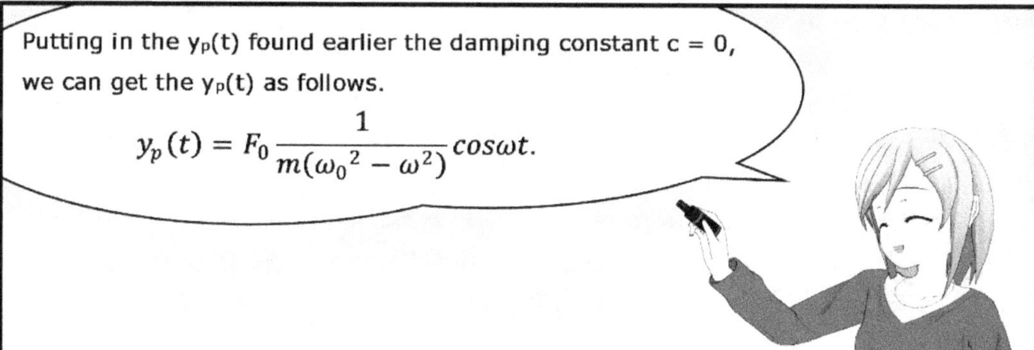

Putting in the yp(t) found earlier the damping constant c = 0, we can get the yp(t) as follows.
$y_p(t) = F_0 \dfrac{1}{m(\omega_0{}^2 - \omega^2)} cos\omega t.$

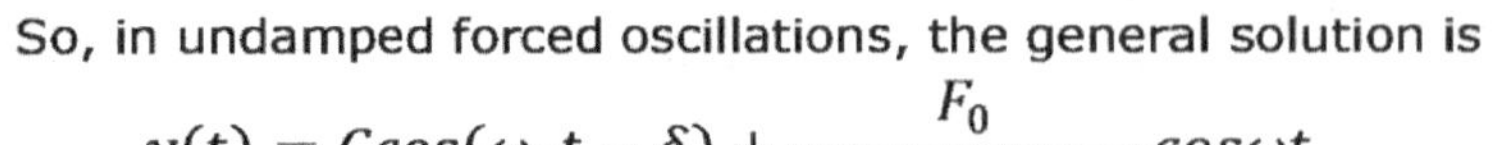

$$y(t) = C\cos(\omega_0 t - \delta) + \frac{F_0}{m(\omega_0{}^2 - \omega^2)}\cos\omega t.$$

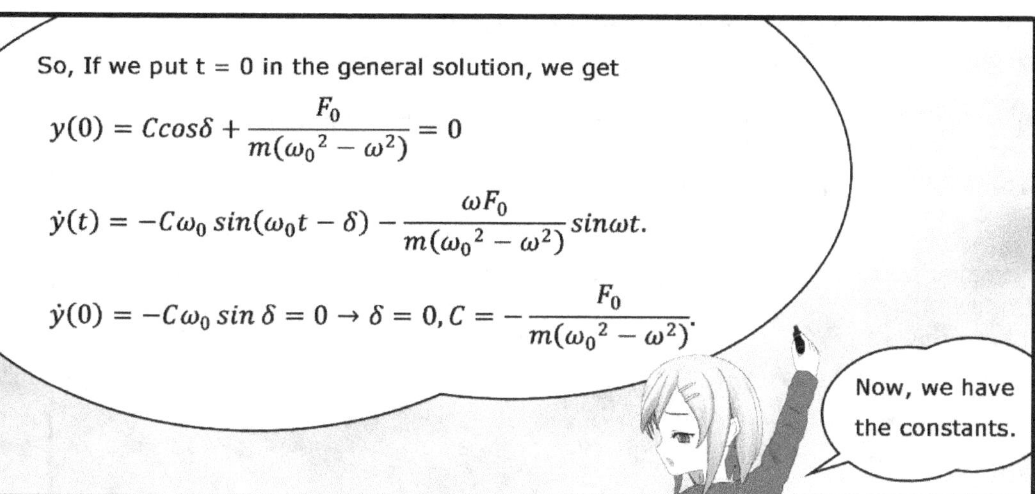

So, If we put t = 0 in the general solution, we get
$$y(0) = C\cos\delta + \frac{F_0}{m(\omega_0^2 - \omega^2)} = 0$$
$$\dot{y}(t) = -C\omega_0 \sin(\omega_0 t - \delta) - \frac{\omega F_0}{m(\omega_0^2 - \omega^2)} \sin\omega t.$$
$$\dot{y}(0) = -C\omega_0 \sin\delta = 0 \rightarrow \delta = 0, C = -\frac{F_0}{m(\omega_0^2 - \omega^2)}.$$
Now, we have the constants.

Therefore, the particular solution satisfying the initial conditions is
$$y(t) = \frac{F_0}{m(\omega_0^2 - \omega^2)}(\cos\omega t - \cos\omega_0 t)$$
$$= \frac{2F_0}{m(\omega_0^2 - \omega^2)} \sin\left(\frac{\omega_0 - \omega}{2} t\right) \sin\left(\frac{\omega_0 + \omega}{2} t\right).$$

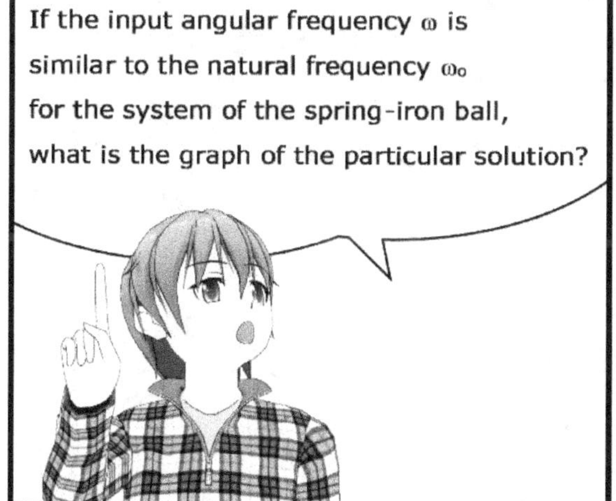

If the input angular frequency ω is similar to the natural frequency ω₀ for the system of the spring-iron ball, what is the graph of the particular solution?

I want to take a break.

Why?
Because I'm tired!

Since $\omega \approx \omega_o$, $\omega - \omega_o$ should be very tiny. Then, the period of $\sin\dfrac{\omega_o - \omega}{2}t$ is awfully large. Since the two functions are multiplied in the solution, the curve will take the form where the curve of $\sin\dfrac{\omega_o + \omega}{2}t$ with smaller period is surrounded by the curve of $\pm\sin\dfrac{\omega_o - \omega}{2}t$ with bigger period.

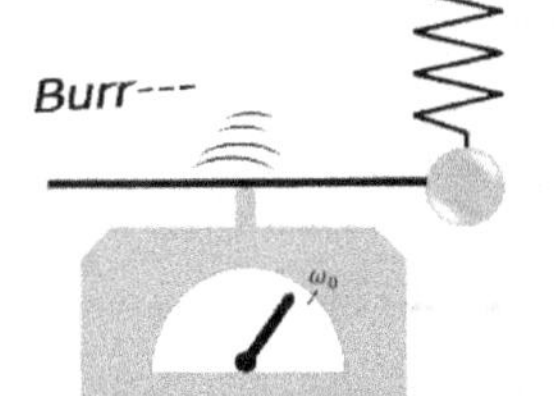

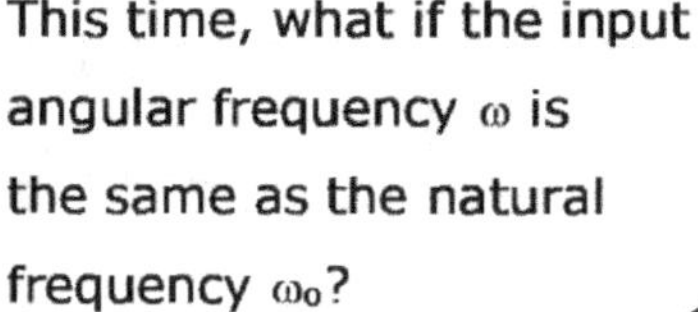

Back to the eq of motion
$$m\ddot{y} + ky = F_0 \cos\omega t.$$
Put
$$\omega = \omega_0 = \sqrt{\frac{k}{m}},$$
into the equation, we get
$$m\ddot{y} + ky = F_0 \cos\omega_0 t.$$

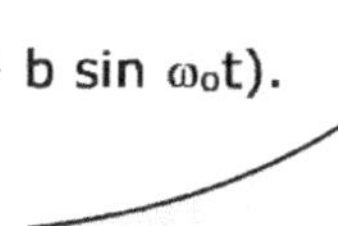

We need to find the undetermined coefficients a and b.

Finding $\ddot{y}_p$ to put it into the new eq above, we get

$$\dot{y}_p = a\cos\omega_0 t + b\sin\omega_0 t + t(-\omega_0 a\sin\omega_0 t + \omega_0 tb\cos\omega_0 t).$$

$$\ddot{y}_p = -\omega_0 a\sin\omega_0 t + \omega_0 b\cos\omega_0 t - \omega_0 a\sin\omega_0 t + \omega_0 b\cos\omega_0 t$$

$$+t(-\omega_0^2 a\cos\omega_0 t - \omega_0^2 b\sin\omega_0 t)$$

$$= -2\omega_0 a\sin\omega_0 t - 2\omega_0 b\cos\omega_0 t + t(-\omega_0^2 a\cos\omega_0 t - \omega_0^2 b\sin\omega_0 t).$$

Putting the $\ddot{y}_p$ and y_p into the new eq, we have

$$-2\omega_0 a\sin\omega_0 t + 2\omega_0 b\cos\omega_0 t + t(-\omega_0^2 a\cos\omega_0 t - \omega_0^2 b\sin\omega_0 t)$$

$$+\omega_0^2\{t(a\cos\omega_0 t + b\sin\omega_0 t)\} = \frac{F_0}{m}\cos\omega_0 t.$$

Consequently

$$-2\omega_0 a = 0, \qquad 2\omega_0 b = \frac{F_0}{m}.$$

Therefore

$$a = 0, \qquad b = \frac{F_0}{2m\omega_0}.$$

So

$$y_p(t) = \frac{F_0}{2m\omega_0} t\sin\omega_0 t.$$

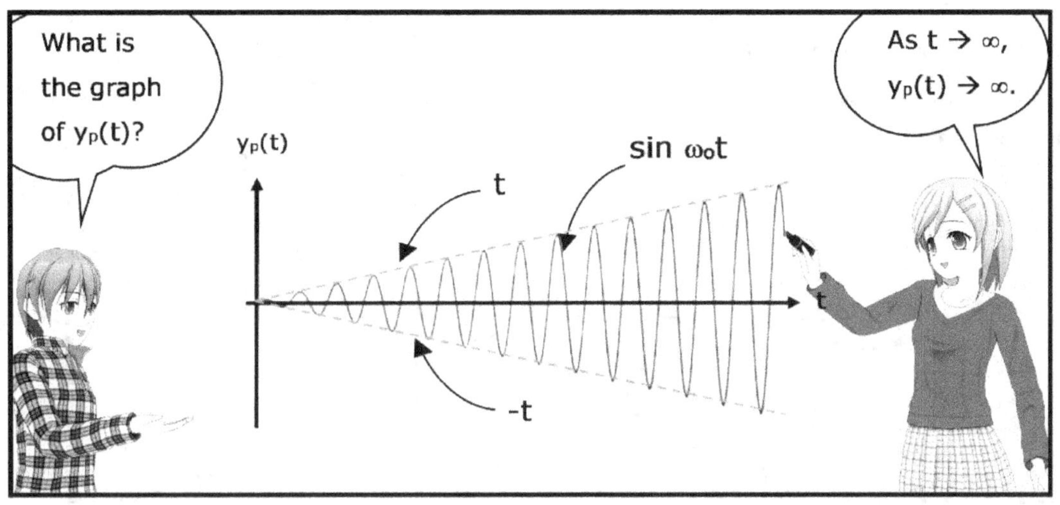
What is
the graph
of yₚ(t)?
yₚ(t)
sin ω₀t
t
-t
t
As t → ∞,
yₚ(t) → ∞.

What is the general
solution, then?

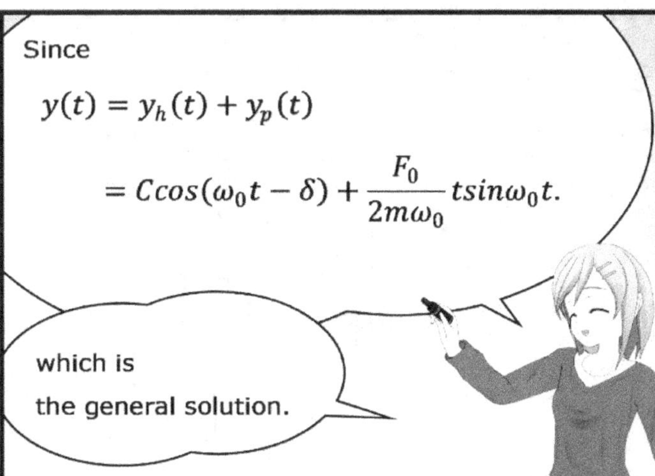
Since
$y(t) = y_h(t) + y_p(t)$
$= C\cos(\omega_0 t - \delta) + \dfrac{F_0}{2m\omega_0} t\sin\omega_0 t.$
which is
the general solution.

As t → ∞, y(t) → ?
Since yₚ(t) → ∞ as t → ∞,
y(t) should go to ∞ also as t → ∞.

The oscillator makes the iron ball oscillate at the constant frequency ω_0 of a small amplitude

This phenomenon is called a sympathetic vibration or a resonance.

If a resonance happens, the mechanical system keeps absorbing energy, and eventually gets destroyed.

Undamped Forced Oscillations

Equation of Motion

$$my'' + ky = F_0 cos\omega_0 t$$

i) General Solution
$(\omega \neq \omega_0)$

$$y(t) = C cos(\omega_0 t - \delta) + \frac{F_0}{m(\omega_0{}^2 - \omega^2)} cos\omega t$$

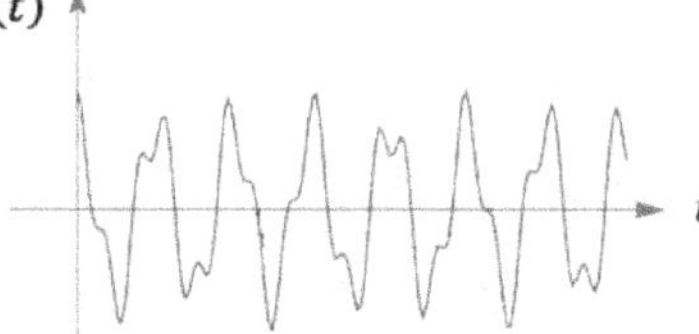

ii) Beat
$(\omega \approx \omega_0)$

$$y(t) = \frac{2F_0}{m(\omega_0{}^2 - \omega^2)} sin\left(\frac{\omega_0 - \omega}{2}t\right) sin\left(\frac{\omega_0 + \omega}{2}t\right)$$

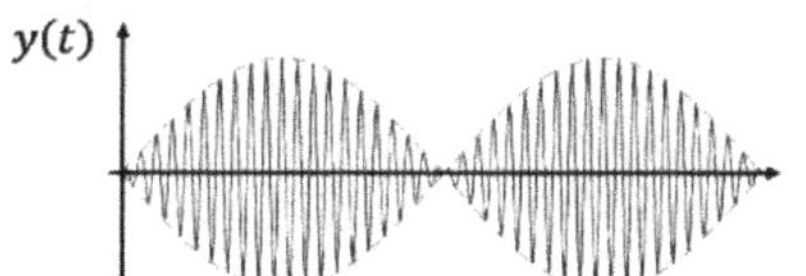

iii) Resonance
$(\omega = \omega_0)$

$$y(t) = C cos(\omega_0 t - \delta) + \frac{F_0}{2m\omega_0} t sin\omega_0 t$$

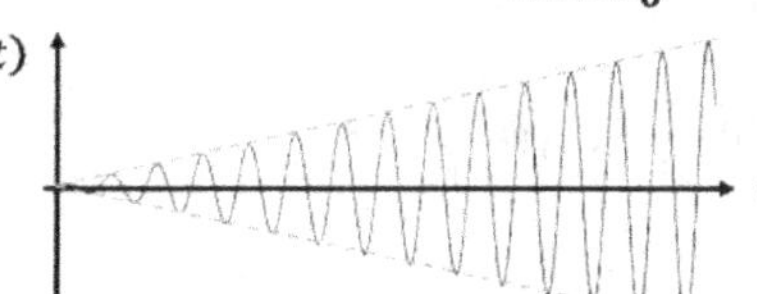

Quiz 1

A vibration system is expressed by diff eq

$$\ddot{y} + 9y = 8cost.$$

Find the particular solution of the system when the initial displacement and velocity is given by $y(0) = 1$ and $\dot{y}(0) = 1$.

Answer

The motion of the system is undamped forced oscillation.

The eq of motion to an undamped forced oscillation is

$$m\ddot{y} + ky = F_0 cos\omega_0 t.$$

Comparing with the given diff eq, we can see $m = 1$, $k = 9$, $F_0 = 8$, $\omega = 1$. And $\omega_0 = \sqrt{k/m} = \sqrt{9/1} = 3$.

Substituting the above values into the general solution of an undamped forced oscillation, we get

$$y(t) = Ccos(\omega_0 t - \delta) + \frac{F_0}{m(\omega_0{}^2 - \omega^2)}cos\omega t$$

$$= C_1 cos\omega_0 t + C_2 sin\omega_0 t + \frac{F_0}{m(\omega_0{}^2 - \omega^2)}cos\omega t$$

$$= C_1 cos3t + C_2 sin3t + cost.$$

Applying the initial conditions, we have

$$y(0) = C_1 cos3(0) + C_2 sin3(0) + cos(0) = C_1 + 1 = 1, C_1 = 0$$

$$\dot{y}(t) = -3C_1 sin3t + 3C_2 cos3t - sint$$

$$\dot{y}(0) = -3C_1 sin3(0) + 3C_2 cos3(0) - sin(0) = 3C_2 = 1, C_2 = \frac{1}{3}.$$

Substituting C_1 and C_2, we finally get

$$y(t) = \frac{1}{3}sin3t + cost.$$

Damped Forced Oscillations
This time, it includes the damping, doesn't it?

Put the dashpot back in.
It's in now.

What is the equation of motion in damped forced oscillations?
It was
$m\ddot{y} + c\dot{y} + ky = F_0 cos\omega t.$

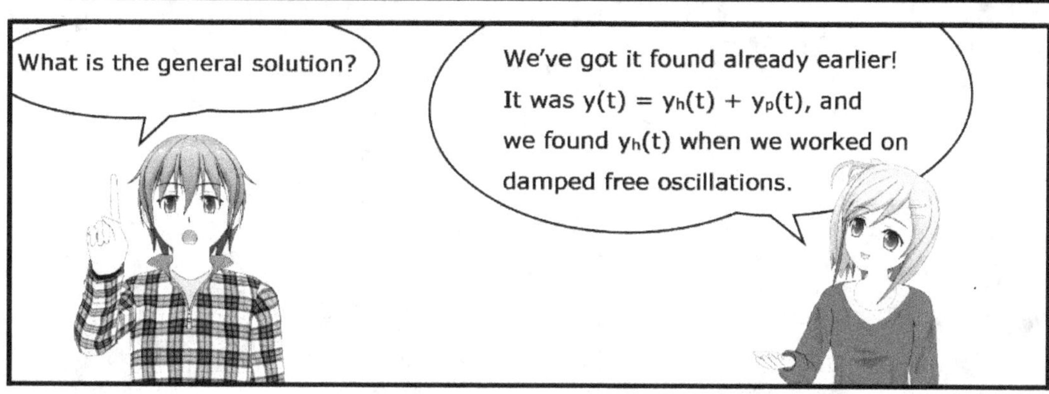

What is the general solution?
We've got it found already earlier! It was y(t) = yₕ(t) + yₚ(t), and we found yₕ(t) when we worked on damped free oscillations.

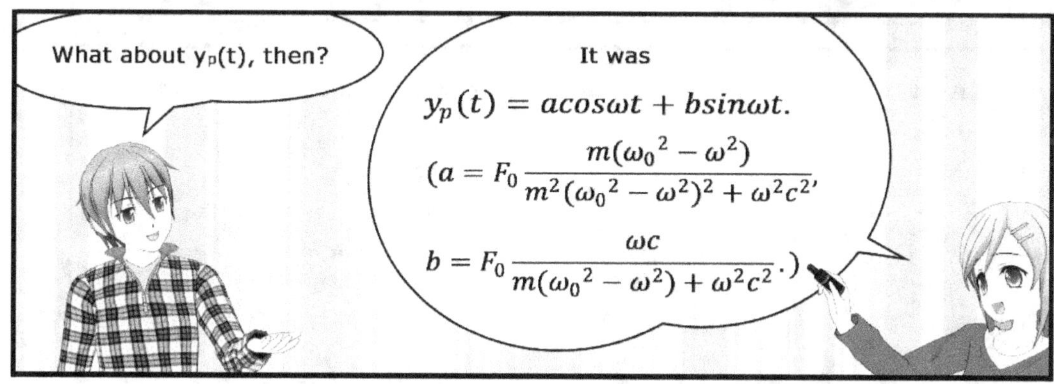

What about yₚ(t), then?
It was
$y_p(t) = acos\omega t + bsin\omega t.$
$(a = F_0 \dfrac{m(\omega_0{}^2 - \omega^2)}{m^2(\omega_0{}^2 - \omega^2)^2 + \omega^2 c^2},$
$b = F_0 \dfrac{\omega c}{m(\omega_0{}^2 - \omega^2) + \omega^2 c^2}.)$

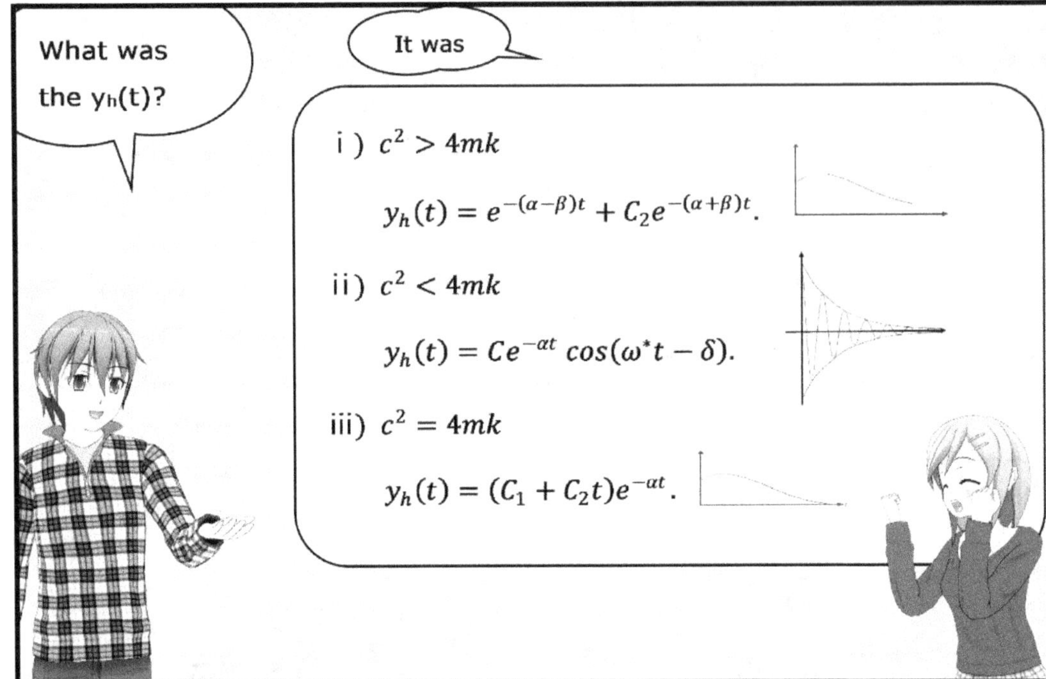

What was the $y_h(t)$?
It was
i) $c^2 > 4mk$
$y_h(t) = e^{-(\alpha-\beta)t} + C_2 e^{-(\alpha+\beta)t}$.
ii) $c^2 < 4mk$
$y_h(t) = Ce^{-\alpha t} cos(\omega^* t - \delta)$.
iii) $c^2 = 4mk$
$y_h(t) = (C_1 + C_2 t)e^{-\alpha t}$.

As t → ∞, $y_h(t)$ → ?
I can see $y_h(t)$ → 0 in all cases. I wonder how!
What are you wondering at? It's natural. A free oscillation means no more external force during the motion, and it is destined to stop due to the existence of damping.
Aha! I see.

Then, what if t → ∞ in damped forced oscillations?
Since $y_h(t)$ → 0, it will eventually converge to $y_p(t)$. In other words, as t → ∞, the iron ball will be oscillating at the input angular frequency ω.

$$y_p(t) = a\cos\omega t + b\sin\omega t$$

$$\left(a = F_0 \frac{m(\omega_0{}^2 - \omega^2)}{m^2(\omega_0{}^2 - \omega^2)^2 + \omega^2 c^2},\right.$$

$$\left.b = F_0 \frac{\omega c}{m(\omega_0{}^2 - \omega^2) + \omega^2 c^2}\right)$$

$$y_p(t) = C^*\cos(\omega t - \eta).$$

$$\left(C^* = \sqrt{a^2 + b^2} = \frac{F_0}{\sqrt{m^2(\omega_0{}^2 - \omega^2)^2 + \omega^2 c^2}}\right.$$

$$\left.\tan\eta = \frac{b}{a} = \frac{\omega c}{m(\omega_0{}^2 - \omega^2)}\right)$$

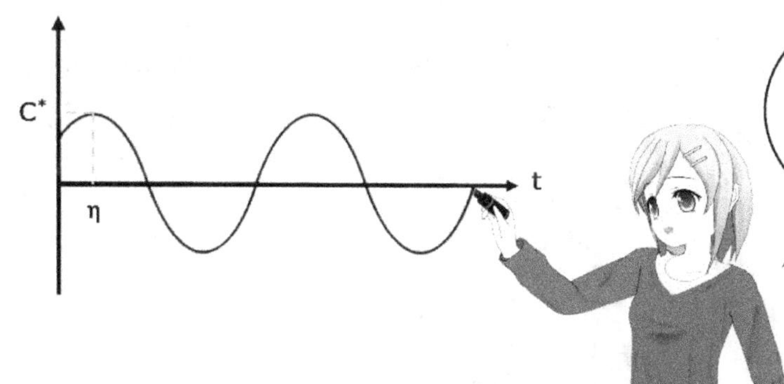

It is. Eventually, the amplitude of the iron ball will be C*
in the steady state. Now, we need to note that
the C* is a function of ω, C*(ω), and therefore,
the amplitude changes with ω.
Construct the graph of C*(ω).

$C^*(\omega) = \dfrac{F_0}{\sqrt{m^2(\omega_0{}^2 - \omega^2)^2 + \omega^2 c^2}}$
C*(ω)
Therefore, it will roughly
look like the one below.
C*max
The amplitude of the iron ball is
the maximum at a particular
input frequency ω!
ω
ωmax

This is another kind
in resonance also.
Then, what is the difference
in resonance between
damped forced oscillation and
undamped forced oscillation?

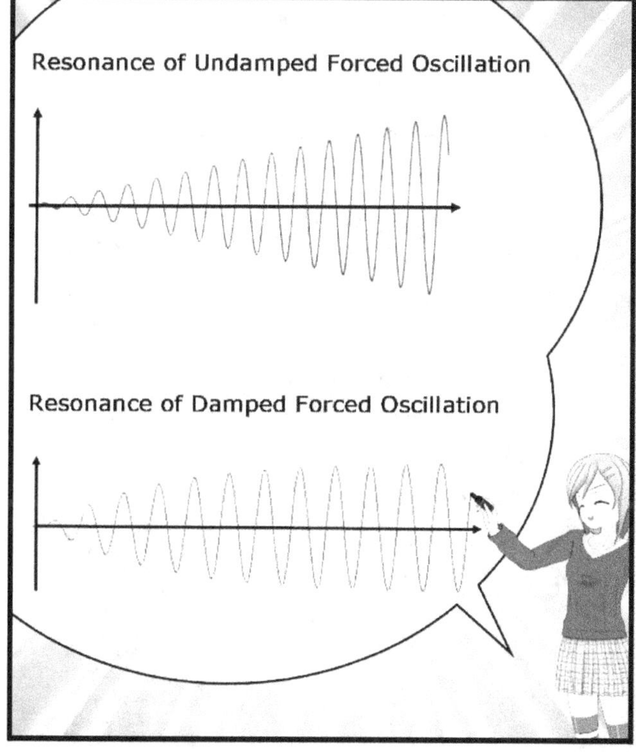

Since every actual
mechanical system is
damped, the amplitude
at the resonance is
limited as in the cases
of damped forced
oscillations. Therefore,
the resonance of damped
forced oscillation is called
the practical resonance.

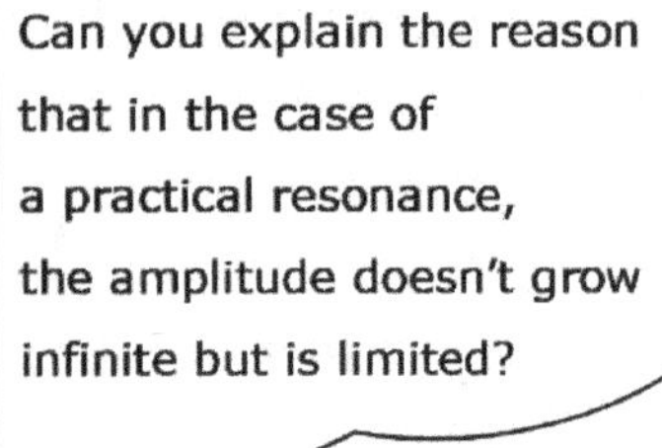

Can you explain the reason that in the case of a practical resonance, the amplitude doesn't grow infinite but is limited?

In that case, the dashpot takes up the external energy to a degree, and therefore, the amplitude may not grow infinite.

Also, the input frequency that causes the practical resonance is not the same as the natural frequency of the mechanical system. Why not?

That's because the dashpot delays activities in the system.

Using the practical resonance, we can amplify the amplitude.

Even though we make the iron ball oscillate with a little amplitude, it gets to oscillate with the larger amplitude if the frequency is close to the frequency that causes the resonance. Such a phenomenon is called the amplification. In other words, the ratio between the amplitude $C^*(\omega)$ with which the iron ball oscillates and the input amplitude F_0 that we use to initiate the iron ball to oscillate is called the amplification.

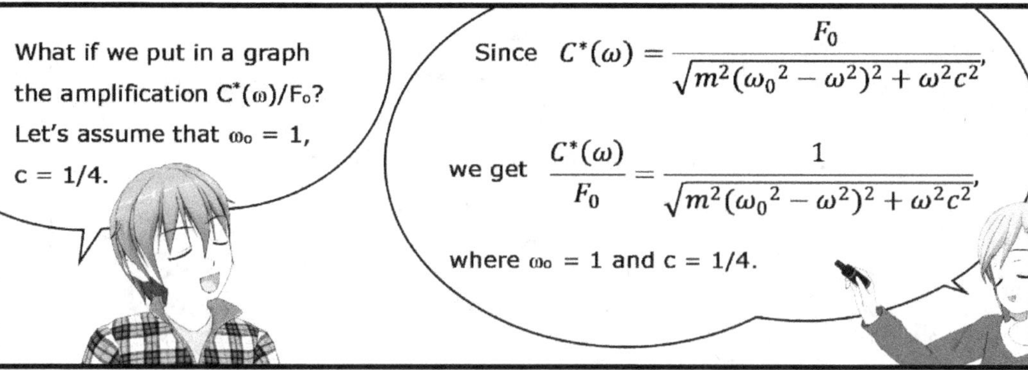

What if we put in a graph the amplification $C^*(\omega)/F_0$? Let's assume that $\omega_0 = 1$, $c = 1/4$.
Since $C^*(\omega) = \dfrac{F_0}{\sqrt{m^2(\omega_0{}^2 - \omega^2)^2 + \omega^2 c^2}}$, we get $\dfrac{C^*(\omega)}{F_0} = \dfrac{1}{\sqrt{m^2(\omega_0{}^2 - \omega^2)^2 + \omega^2 c^2}}$, where $\omega_0 = 1$ and $c = 1/4$.

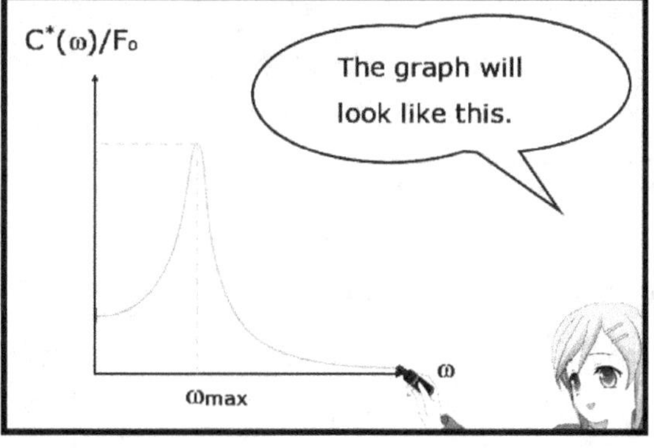

$C^*(\omega)/F_0$
ω
ωmax
The graph will look like this.

It is similar to the one for $C^*(\omega)$ previously constructed.

It is natural since a division of $C^*(\omega)$ by F_0 is the only difference.
What graph do we get if we choose for the damping constant c other values instead of 1/4?

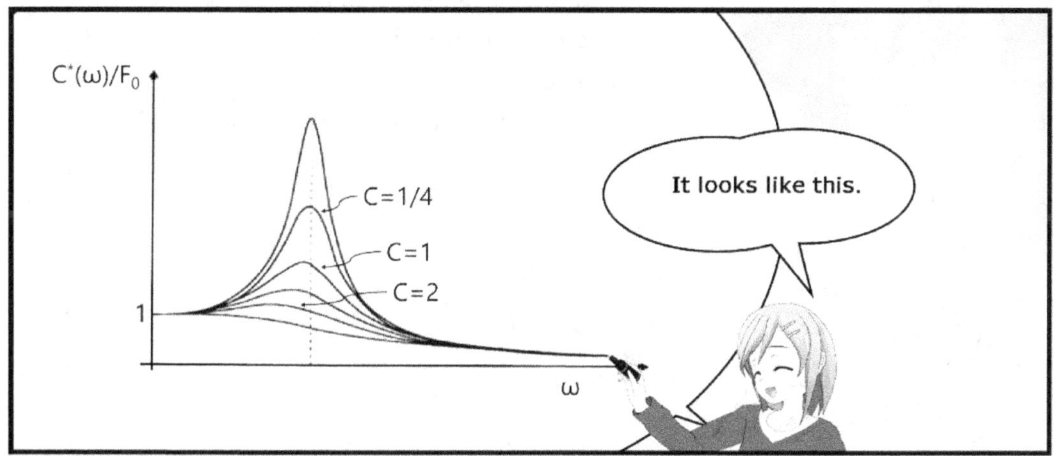

$C^*(\omega)/F_0$
ω
C=1/4
C=1
C=2
1
It looks like this.

What change will result if the damping constant c gets larger?
The degree of amplification gets smaller. Consequently, the amplitude at the practical resonance gets smaller also. Come to think of it, it is natural that the amplitude gets smaller as the damping gets larger.

Is that it?
The frequency at the practical resonance gets smaller too. Besides, if the damping constant is too big, the practical resonance doesn't occur.

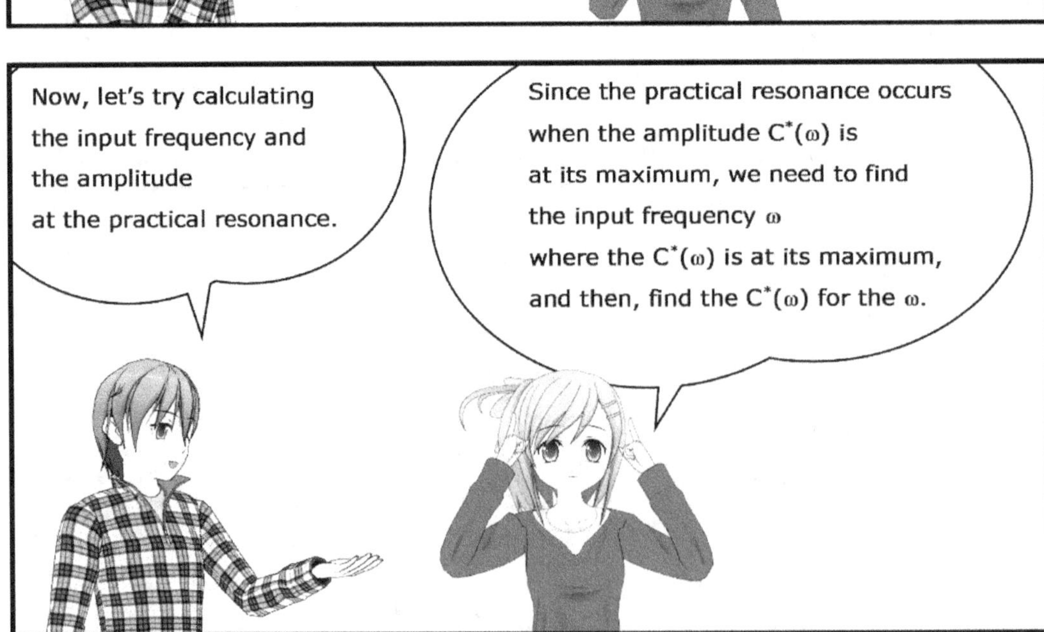

Now, let's try calculating the input frequency and the amplitude at the practical resonance.
Since the practical resonance occurs when the amplitude $C^*(\omega)$ is at its maximum, we need to find the input frequency ω where the $C^*(\omega)$ is at its maximum, and then, find the $C^*(\omega)$ for the ω.

$$\frac{dC^*(\omega)}{d\omega} = F_0\left(-\frac{1}{2}\right)\underbrace{\{m^2(\omega_0{}^2 - \omega^2)^2 + \omega^2 c^2\}^{-\frac{3}{2}}\{-4m^2\omega(\omega_0{}^2 - \omega^2) + 2\omega c^2\}}_{0}.$$

Therefore

$$-4m^2\omega(\omega_0{}^2 - \omega^2) + 2\omega c^2 = 0.$$

$$\omega^2 = \frac{2m^2\omega_0^2 - c^2}{2m^2}.$$

Practical Resonance Laboratory
The last section in Mechanics Pavilion!
WARNING
No Children and Pregnant women are allowed.
Office of Management

Let's set the index of the earthquake generator to ω_{max} the frequency at the practical resonance for this building.
Burr---
ω_{max}

Ugh------
Earthquake!
It's shaking harder now.
Yipe! The building's falling!

See!
Even if the earthquake is weak,
a building fails to hold and
gets destroyed if the frequency is
the same as the frequency
at the practical resonance
for the building.

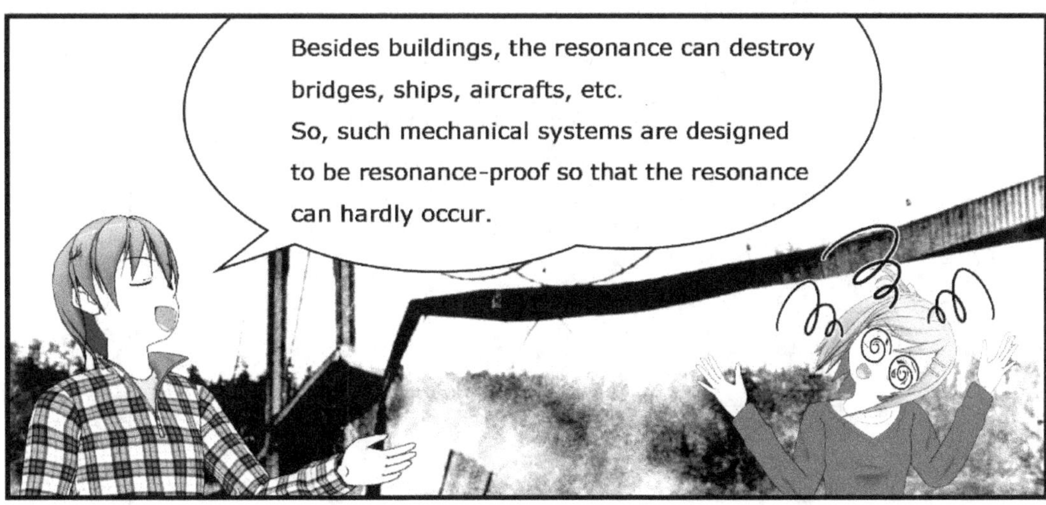

Besides buildings, the resonance can destroy
bridges, ships, aircrafts, etc.
So, such mechanical systems are designed
to be resonance-proof so that the resonance
can hardly occur.

Congratulations on
your passing
the Mechanics Pavilion!
Now, let's move on
to the last course,
Electricity Pavilion.

Damped Forced Oscillations

Equation of Motion

$$m\ddot{y} + c\dot{y} + ky = F_0\cos\omega t$$

General Solution $\quad y(t) = y_h(t) + y_p(t)$

$$y_p(t) = C^*\cos(\omega t - \eta).$$

$$\left(C^* = \sqrt{a^2 + b^2} = \frac{F_0}{\sqrt{m^2(\omega_0{}^2 - \omega^2)^2 + \omega^2 c^2}}\right.$$

$$\left.\tan\eta = \frac{b}{a} = \frac{\omega c}{m(\omega_0{}^2 - \omega^2)}\right)$$

Practical Resonance

$$\omega = \sqrt{\frac{2m^2\omega_0{}^2 - c^2}{2m^2}} \qquad C^* = \frac{2mF_0}{c\sqrt{4m^2\omega_0^2 - c^2}}$$

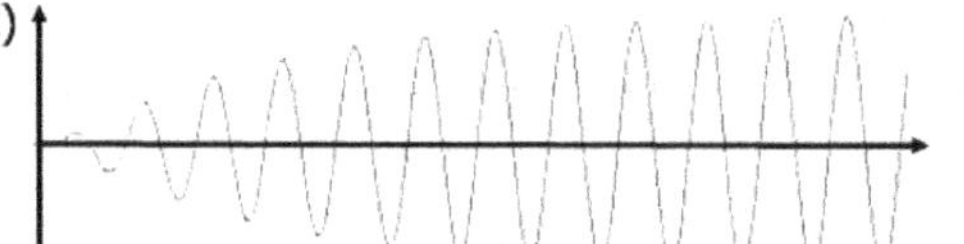

Quiz 1

Find the transient and steady-state motions of the vibrating system governed by the eq

$$\ddot{y} + 3\dot{y} + 2y = cost.$$

Answer

The general solution to the nonhomogeneous diff-eq is

$y(t) = y_h(t) + y_p(t)$ where y_h is the general solution

to the homogeneous diff-eq and for the transient motions

and y_p is the particular solution to the nonhomogeneous diff-eq

and for the steady-state motions.

i) Since y_h is the solution to

$$\ddot{y} + 3\dot{y} + 2y = 0,$$

the characteristic eq is

$$\lambda^2 + 3\lambda + 2 = 0, \qquad (\lambda + 1)(\lambda + 2) = 0, \qquad \lambda = -1, -2.$$

Therefore, the transient motions are given by

$$y_h = C_1 e^{-t} + C_2 e^{-2t}.$$

ii) Since y_p obtained above was

$$y_p = a\cos\omega t + b\sin\omega t$$

where

$$\left(a = F_0 \frac{m(\omega_0^2 - \omega^2)}{m^2(\omega_0^2 - \omega^2)^2 + \omega^2 c^2}, \qquad b = F_0 \frac{\omega c}{m(\omega_0^2 - \omega^2) + \omega^2 c^2}\right)$$

Comparing to the eq of motion

$$m\ddot{y} + c\dot{y} + ky = F_0\cos\omega t,$$

we see that $m = 1$, $c = 3$, $k = 2$, $F_0 = 1$, $\omega = 1$, and $\omega_0 = \sqrt{k/m} = \sqrt{2}.$

Substituting these value, we get $a = 1/10$ and $b = 3/10$.

Therefore

$$y_p = \frac{1}{10} e^{-t} + \frac{3}{10} e^{-2t}.$$

Quiz 2

What do we need to consider maximizing the sensitivity of a seismograph?

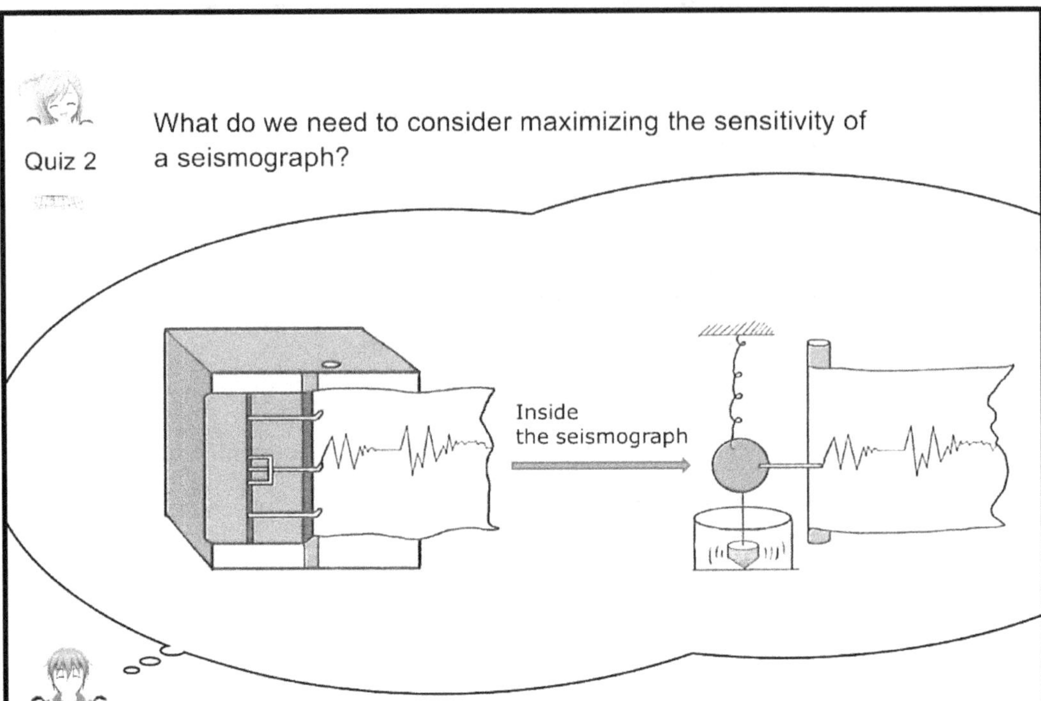

Answer

We can simplify a seismograph by means of a damped forced oscillation model where the equation of motion is

$$m\ddot{y} + c\dot{y} + ky = F_0 \cos\omega t.$$

where the ω is the frequency of the earthquake vibrating the seismograph.

Now, when

$$\omega = \sqrt{\frac{2m^2\omega_0{}^2 - c^2}{2m^2}} = \sqrt{\frac{k}{m} - \frac{c^2}{2m^2}} \qquad (\because \omega_0 = \sqrt{\frac{k}{m}})$$

the practical resonance occurs and the seismograph vibrates at its maximum.

Therefore, we take many measurements of the frequencies of earthquakes and take the average of the measurements, and when we design a seismograph, we set the values of m, k, and c appropriately and make the value of

$$\sqrt{\frac{k}{m} - \frac{C^2}{2m^2}}$$ of the seismograph equal to the average.

Quiz 3

We have considered the resonance for the object that oscillates; however, the resonance occurs for an object that rotates also. Make two examples of resonance for rotating objects.

Answer

1) A spinning top

When we slap a top by hand to make it spin, it spins fastest if the slapping frequency is the same as the frequency at the practical resonance for the top. That's because the slapping energy gets absorbed into the top if the practical resonance occurs. Just slapping fast doesn't make the top spin faster.

2) Magnetic Resonance Imaging (MRI)

MRI is used when we diagnose brain diseases taking photos of cross sections of a brain. Applying AC magnetic field to trigger the practical resonance of spinning motions of atomic nuclei of particular material composing a brain, we can see how the material is distributed in the brain since the energy gets absorbed in the places where the material exist. For instance, since the distributions of the material in a normal brain and a brain with tumors are different, we can see if the brain has tumors.

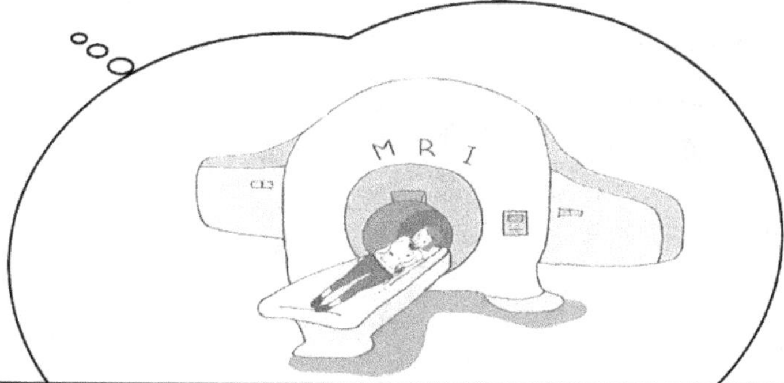

Applications of 2nd Order Nonhomogeneous Diff-eqs with Const-coeffs II

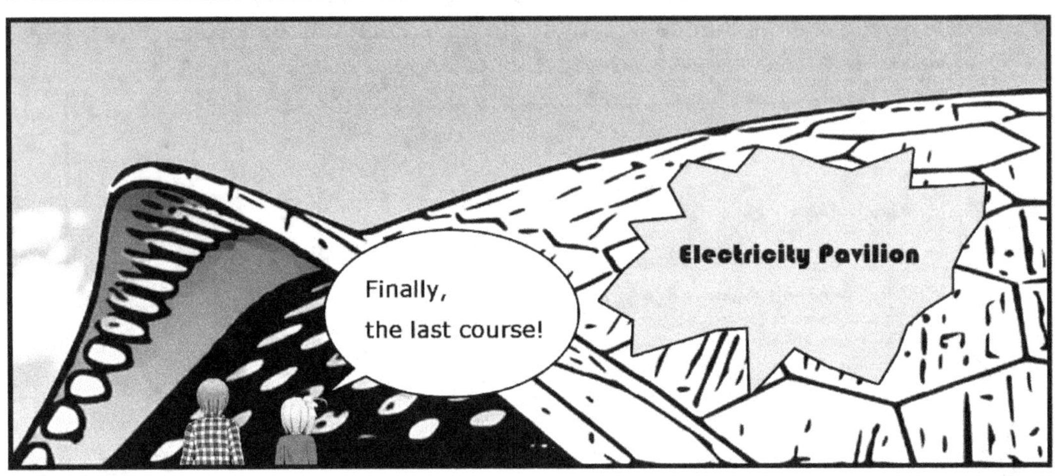

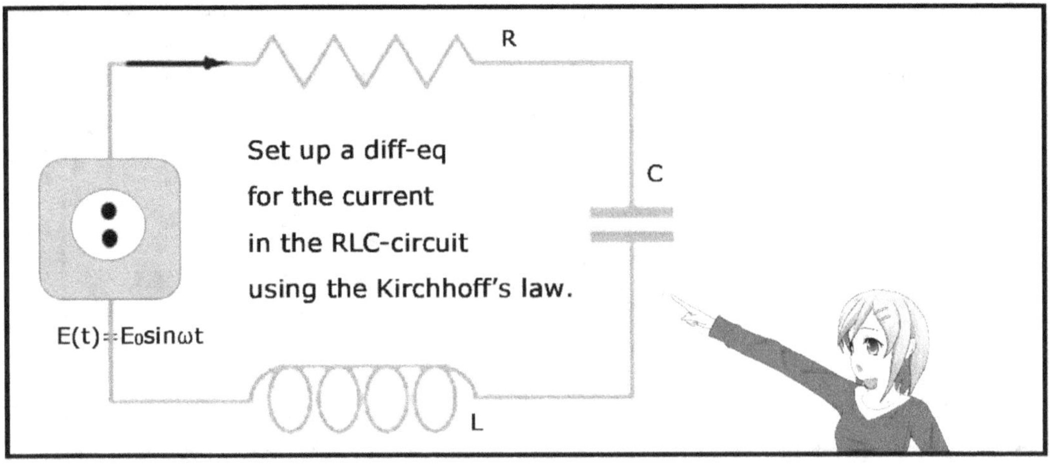

What was the Kirchhoff's law?
It was the fact that in a closed loop, the voltage impressed by the power source is the same as the sum of the voltage drops in the rest of the loop.

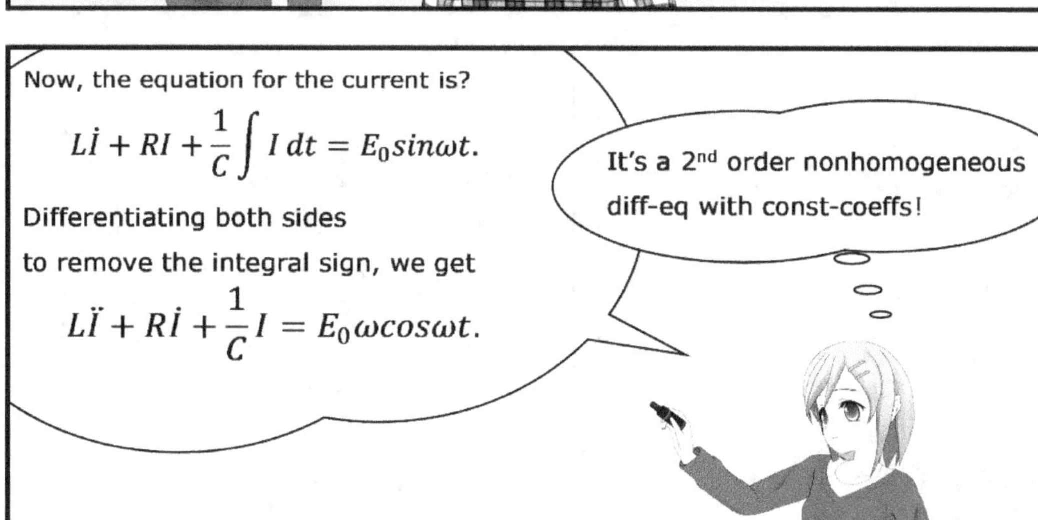

Now, the equation for the current is?
$$Li + RI + \frac{1}{C}\int I\, dt = E_0 \sin\omega t.$$
Differentiating both sides to remove the integral sign, we get
$$L\ddot{I} + R\dot{I} + \frac{1}{C}I = E_0\,\omega\cos\omega t.$$
It's a 2nd order nonhomogeneous diff-eq with const-coeffs!

Isn't that surprising?
What is it?

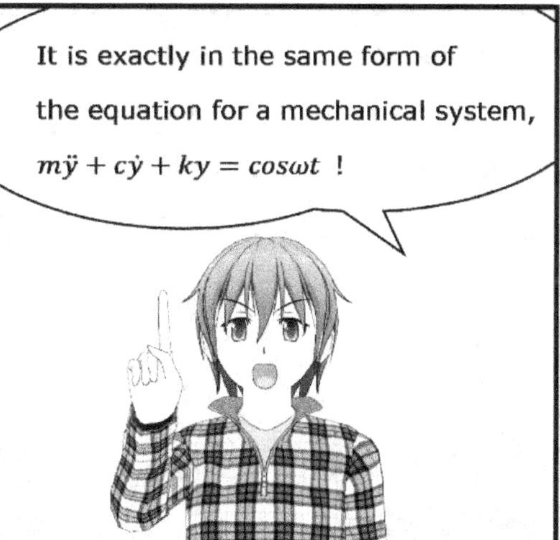

It is exactly in the same form of the equation for a mechanical system,
$$m\ddot{y} + c\dot{y} + ky = \cos\omega t \ !$$

What do we say for this?
The greatness of math!
Despite completely different phenomena, if we put them mathematically, we end up with the same unified expression!

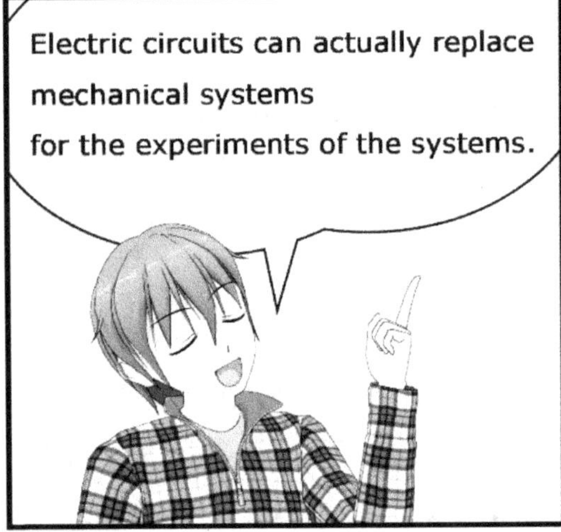

Electric circuits can actually replace mechanical systems for the experiments of the systems.

By such replacements, we don't have to have headaches due to troublesomeness making the structures or inaccuracy of the measurements.

That's because electric circuits can be made simply and the measurements are easy.
Aha!

Comparison between Mechanical Systems and Electric Systems

Mechanical Systems

$$m\ddot{y} + c\dot{y} + ky = F_0 cos\omega t$$

m: Mass

c: Damping Constant

k: Spring Constant

$F_0 cos\omega t$: Force Applied

y(t): Displacement

Electric Systems

$$L\ddot{I} + R\dot{I} + \frac{1}{C}I = E_0\omega cos\omega t$$

L: Inductance

R: Resistance

1/C: Inverse of Capacitance

$E_0\omega cos\omega t$: Derivative of Voltage Input

I(t): Current

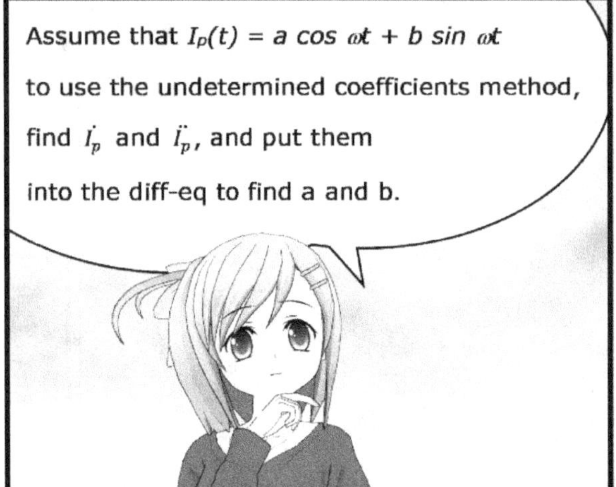

Assume that $I_P(t) = a\cos\omega t + b\sin\omega t$ to use the undetermined coefficients method, find $\dot{I}_p$ and $\ddot{I}_p$, and put them into the diff-eq to find a and b.
Omitting the calculation details and putting the result only, we get
$$a = \frac{-E_0 S}{R^2 + S^2},$$
$$b = \frac{-E_0 R}{R^2 + S^2}.$$
$$(S = \omega L - \frac{1}{\omega C})$$

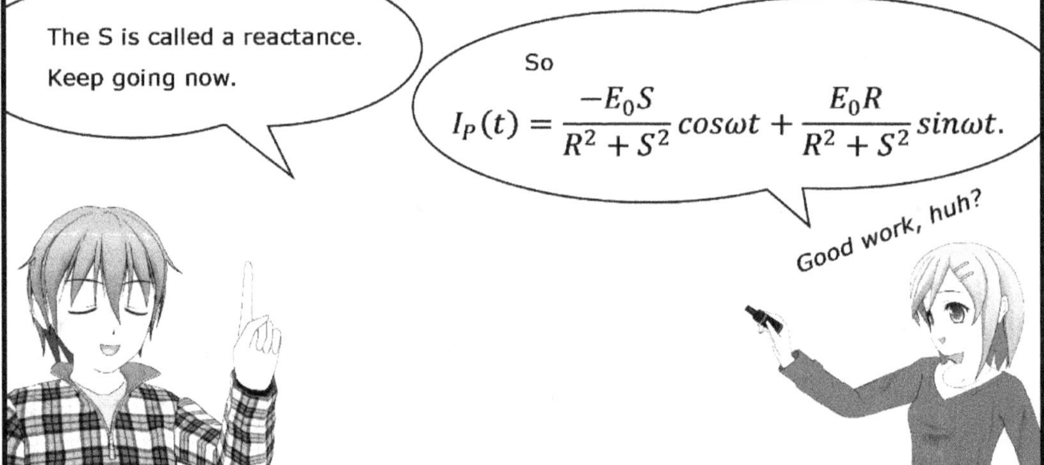

The S is called a reactance. Keep going now.
So
$$I_P(t) = \frac{-E_0 S}{R^2 + S^2}\cos\omega t + \frac{E_0 R}{R^2 + S^2}\sin\omega t.$$
Good work, huh?

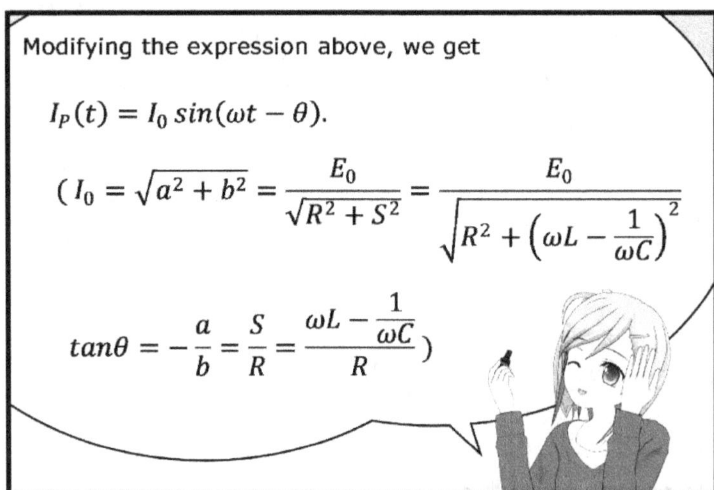

Modifying the expression above, we get
$$I_P(t) = I_0 \sin(\omega t - \theta).$$
$$(I_0 = \sqrt{a^2 + b^2} = \frac{E_0}{\sqrt{R^2 + S^2}} = \frac{E_0}{\sqrt{R^2 + \left(\omega L - \frac{1}{\omega C}\right)^2}}$$
$$\tan\theta = -\frac{a}{b} = \frac{S}{R} = \frac{\omega L - \frac{1}{\omega C}}{R})$$

$$\sqrt{R^2 + \left(\omega L - \frac{1}{\omega C}\right)^2}$$
is called impedance.

Impedance is pretty close to kickboxing, frying pan, and Kentucky Fried Chicken.

Why?
Sputters a lot when it's said!

Impedance is a collection of all the fellows impeding the flow of current. Taking Z for impedance, we get $I_0 = E_0/Z$.
It is in the same form of the one in Ohm's law, I = E/R.

Does a resonance occur in an RLC-circuit too?

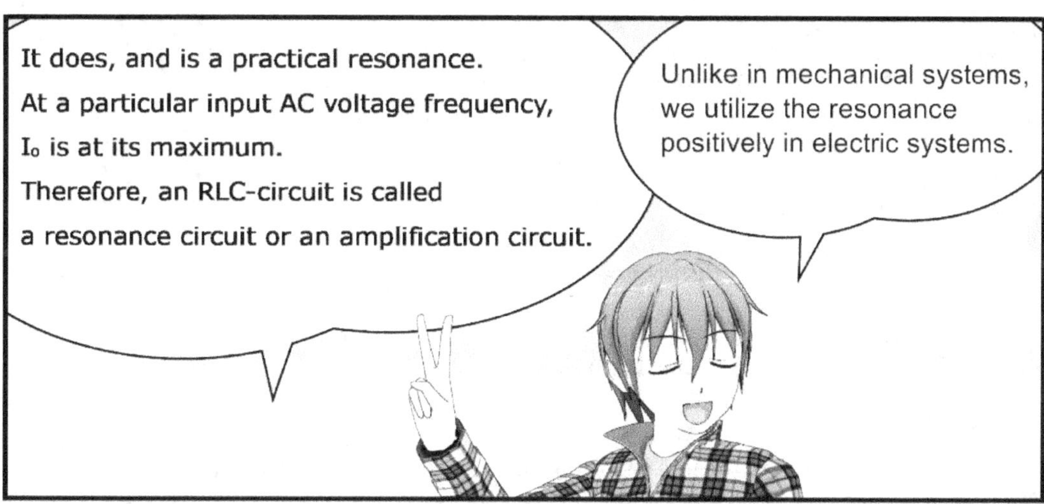

It does, and is a practical resonance. At a particular input AC voltage frequency, I_0 is at its maximum. Therefore, an RLC-circuit is called a resonance circuit or an amplification circuit.
Unlike in mechanical systems, we utilize the resonance positively in electric systems.

Quiz 1

Find the general solution for the current $I(t)$ of an RLC circuit with resistance $R=100$ Ω, inductance $L=0.1$ H, capacitance $C= 1$ mF, which is connected to a source of voltage $E(t)=10\ sin500t$.

Answer

Applying Kirchhoff's law, we get $L\dot{I} + RI + \dfrac{1}{C}\displaystyle\int I\,dt = E_0 sin\omega t.$

Differentiating both sides, we have $L\ddot{I} + R\dot{I} + \dfrac{1}{C}I = E_0\omega cos\omega t.$

$$E_0 = 10, \quad \omega = 500$$

Substituting the given values, we get $0.1\ddot{I} + 100\dot{I} + 1000I = 5000cos500t. \cdots (*)$

Since it is a nonhomogeneous diff-eq, the general solution is

$$I(t) = I_h(t) + I_p(t),$$

where I_h is the general solution to the homogeneous diff-eq and I_p is the particular solution to the nonhomogeneous diff-eq.

i) Since I_h is the general solution to

$$0.1\ddot{I} + 100\dot{I} + 1000I = 0,$$

the characteristic eq is

$$0.1\lambda^2 + 100\lambda + 1000 = 0.$$

Therefore

$$\lambda = \frac{-100 \pm \sqrt{100^2 - 4(0.1)(1000)}}{2(0.1)} \cong \frac{-100 \pm 98}{0.2} = \begin{cases} -10 \\ -990 \end{cases}.$$

$$I_h(t) = C_1 e^{-10t} + C_2 e^{-990t}$$

ii) Since I_p obtained above was

$$I_p(t) = \frac{-E_0 S}{R^2 + S^2} cos\omega t + \frac{E_0 R}{R^2 + S^2} sin\omega t \qquad \left(S = \omega L - \frac{1}{\omega C}\right).$$

$$S = 48$$

Substituting the given values, we get

$$\frac{-E_0 S}{R^2 + S^2} = \frac{-(10)(48)}{(100)^2 + (48)^2} = -0.039, \qquad \frac{E_0 R}{R^2 + S^2} = \frac{(10)(100)}{(100)^2 + (48)^2} = 0.081.$$

So

$$I_p(t) = -0.039cos500t + 0.081sin500t.$$

Therefore

$$I(t) = I_h(t) + I_p(t) = C_1 e^{-10t} + C_2 e^{-990t} - 0.039cos500t + 0.081sin500t$$

Quiz 2

Answer

What do we mean by tuning a radio channel to listen to
a certain station, after all?

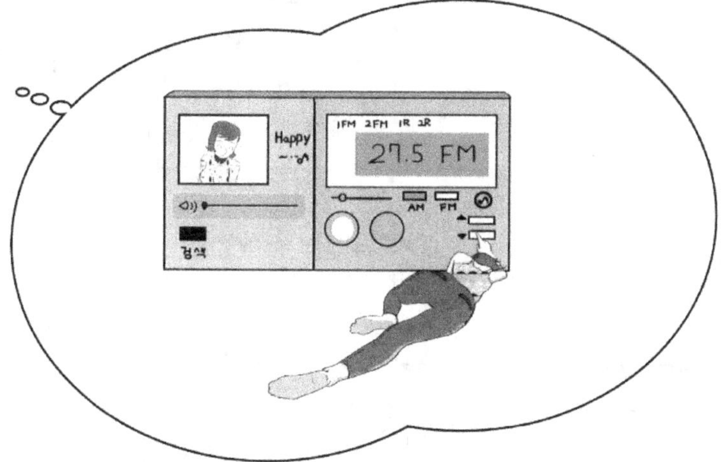

Since we can take a radio for an RLC circuit, the tuning means
maximizing (practical resonance) the amplitude of
the steady state current by adjusting the capacitance C.
Let's now find the capacitance C at the practical resonance.

The practical resonance occurs when the amplitude I_0 of

the steady state current $I_p(t) = I_0 sin(\omega t - \theta)$ is at its maximum.

Now, since

$$I_0(\omega) = \frac{E_0}{\sqrt{R^2 + \left(\omega L - \frac{1}{\omega C}\right)^2}},$$

the practical resonance occurs when $dI_0(\omega)/d\omega = 0$.

$$\frac{dI_0(\omega)}{d\omega} = E_0\left(-\frac{1}{2}\right)\left[R^2 + \left(\omega L - \frac{1}{\omega C}\right)^2\right]^{-\frac{3}{2}} 2\left(\omega L^2 - \frac{1}{\omega^3 C^2}\right) = 0.$$

Therefore, if we set

$$\omega L^2 = \frac{1}{\omega^3 C^2}, \qquad C = \frac{1}{\omega^2 L}$$

we get the tuning done.

Applications of Systems of Diff-eqs

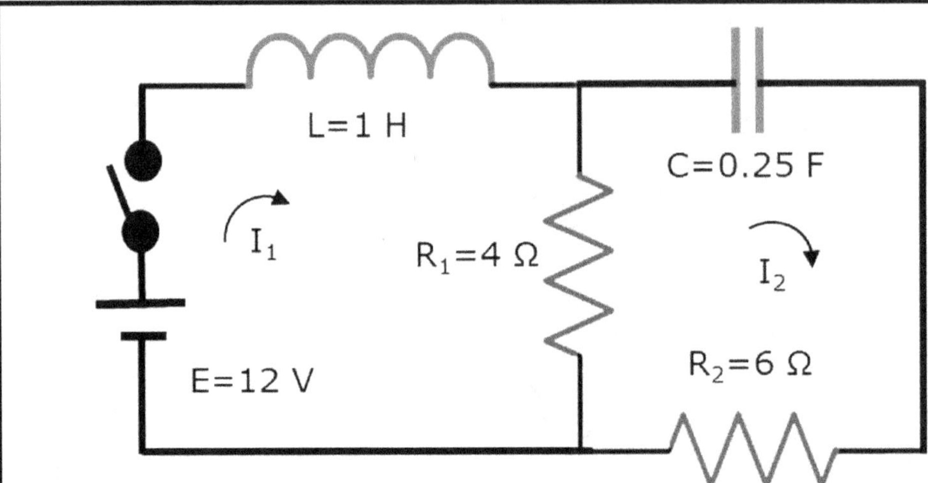

Applying the Kirchhoff's law, for the left loop, we get

$$L\dot{I_1} + R_1(I_1 - I_2) = E. \quad \text{(The current through } R_1 \text{ is } I_1 - I_2.)$$

For the right loop, we get

$$R_2 I_2 + R_1(I_2 - I_1) + \frac{1}{C}\int I_2\, dt = 0.$$

Differentiating both sides, we get

$$R_2 \dot{I_2} + R_1(\dot{I_2} - \dot{I_1}) + \frac{1}{C} I_2 = 0.$$

Putting in the system of eqs the values given and simplifying the eqs, we get

$$\dot{I}_1 + 4I_1 - 4I_2 = 12 \qquad (1)$$

$$-4\dot{I}_1 + 10\dot{I}_2 + 4I_2 = 0 \qquad (2)$$

Keep going.

Modifying the eq (1) to eliminate the I_2, we get

$$I_2 = \frac{1}{4}\dot{I}_1 + I_1 - 3.$$

Differentiating both sides, we get

$$\dot{I}_2 = \frac{1}{4}\ddot{I}_1 + \dot{I}_1.$$

Putting I_2 and $\dot{I}_2$ into the eq (2), we get

$$-4\dot{I}_1 + 10\left(\frac{1}{4}\ddot{I}_1 + \dot{I}_1\right) + 4\left(\frac{1}{4}\dot{I}_1 + I_1 - 3\right) = 0.$$

Simplifying the eq above, we get

$$\ddot{I}_1 + \frac{14}{5}\dot{I}_1 + \frac{8}{5}I_1 = \frac{24}{5}.$$

Since it is a 2nd order nonhomogeneous diff-eq with const-coeffs, the general solution should be

$$I_1(t) = I_{1h} + I_{1p}.$$

Good! Keep going.

i) Let's find I_{1h}, first. The I_{1h} is the general solution to the homogeneous eq below.

$$\ddot{I}_1 + \frac{14}{5}\dot{I}_1 + \frac{8}{5}I_1 = 0.$$

Therefore, the characteristic eq is as follows.

$$\lambda^2 + \frac{14}{5}\lambda + \frac{8}{5} = 0.$$

So

$$\lambda_{1,2} = \frac{-\frac{14}{5} \pm \sqrt{\left(\frac{14}{5}\right)^2 - \frac{32}{5}}}{2} = \begin{cases} -0.8 \\ -2 \end{cases}.$$

Therefore

$$I_{1h} = C_1 e^{-2t} + C_2 e^{-0.8t}.$$

ii) Next, let's find the I_{1p} now. The right hand side of the nonhomogeneous eq given is 24/5, which reminds us of the method of undetermined coefficients. So, let's make an assumption as follows.

$$I_{1p} = k.$$

To find the undetermined coefficient k, put in the eq given the expressions below.

$$\dot{I}_{1p} = \ddot{I}_{1p} = 0, \qquad I_{1p} = k,$$

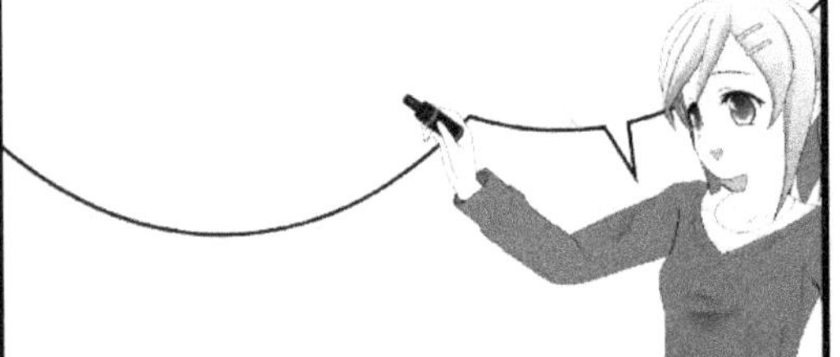

Then, we get the k as shown below.

$$\frac{8}{5}k = \frac{24}{5}. \quad k = 3.$$

So

$$I_{1p} = 3.$$

Therefore, the general solution is

$$I_1(t) = C_1 e^{-2t} + C_2 e^{-0.8t} + 3.$$

$$I_2 = \frac{1}{4}\dot{I}_1 + I_1 - 3.$$

Therefore

$$I_1 = C_1 e^{-2t} + C_2 e^{-0.8t} + 3.$$

$$\dot{I}_1 = -2C_1 e^{-2t} - 0.8C_2 e^{-0.8t}.$$

$$I_2 = \frac{1}{4}(-2C_1 e^{-2t} - 0.8C_2 e^{-0.8t})$$

$$+(C_1 e^{-2t} + C_2 e^{-0.8t} + 3) - 3$$

$$= \frac{1}{2}C_1 e^{-2t} + \frac{4}{5}C_2 e^{-0.8t}.$$

$$I_1(0) = 0,\ I_2(0) = 0.$$

Therefore

$$I_1(0) = C_1 + C_2 + 3 = 0.$$

$$I_2(0) = \frac{1}{2}C_1 + \frac{4}{5}C_2 = 0.$$

$$C_1 = -8,\ C_2 = 5.$$

So, the particular solution satisfying the initial conditions is

$$I_1 = -8e^{-2t} + 5e^{-0.8t} + 3,$$

$$I_2 = -4e^{-2t} + 4e^{-0.8t}.$$

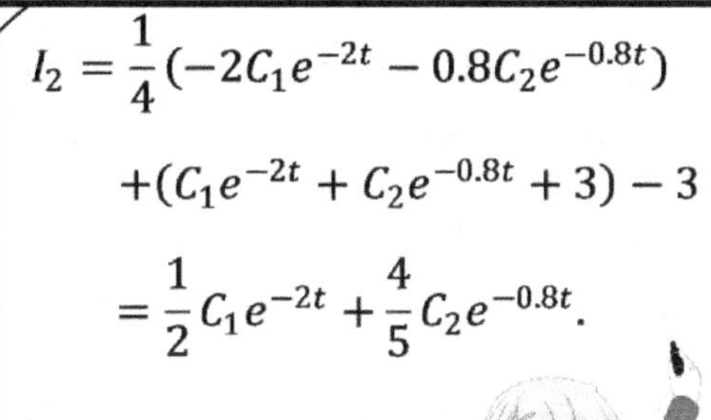

We need to consider the two things below
if we want to make a model for a physical actuality.

1. A series of local laws governing various physical quantities
- They are put in expressions in terms of diff-eqs as usual.

2. Boundary conditions including a series of initial conditions
- They let us calculate to find what will happen to other part of
the universe by indicating the state of a part of the universe
at a particular moment of time.

Stephen Hawking

Quiz 1

Answer

Find the equation of motion of the mechanical system made up of two springs and two iron balls as shown in the figure below.

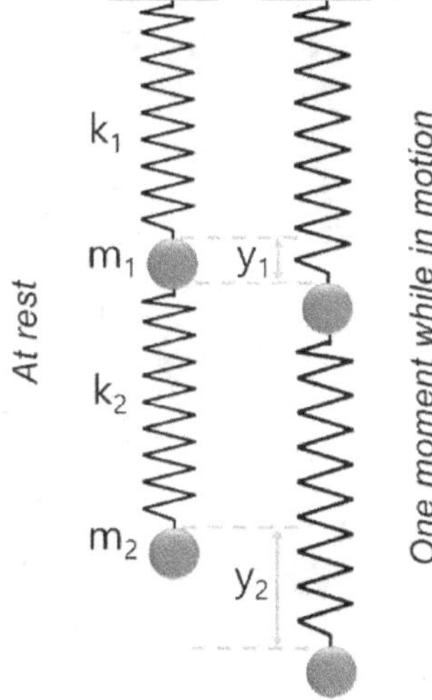

Calculate the forces on the two iron balls
to apply the Newton's second law.

The force on the iron ball 1
    The force upward due to the spring 1: $-k_1 y_1$
    The force downward due to the spring 2: $k_2(y_2 - y_1)$
Therefore, the eq of motion of the iron ball 1 is

$$m_1 \ddot{y}_1 = -k_1 y_1 + k_2(y_2 - y_1).$$

The force on the iron ball 2
    The force upward due to the spring 2: $-k_2(y_2 - y_1)$
Therefore, the eq of motion of the iron ball 2 is

$$m_2 \ddot{y}_2 = -k_2(y_2 - y_1).$$

We now have a system of two diff-eqs.

Quiz 2

Two tanks are connected with the pipes as shown in the figure below. Initially, the tank 1 has 20 liters of water containing 150 grams of sugar dissolved and the tank 2 has 10 liters of water having 50 grams of sugar dissolved. Now, at the respective velocities described in the figure, the two sugar waters in the tanks are getting mixed through the pipes while the tank 1 is getting pure water and the two tanks are losing their liquids. Now, set up a diff-eq for the amount of sugar in each tank at time t.

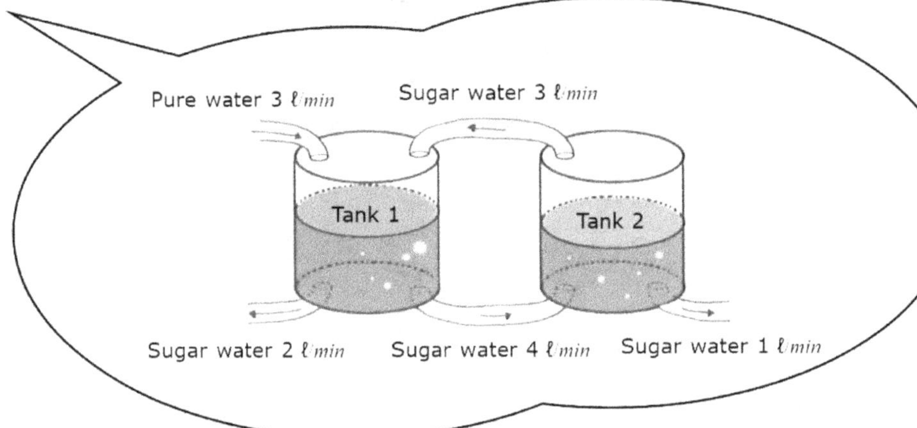

Answer

The amounts of liquid in both tanks remain the same as their initial amounts of 20 liters and 10 liters respectively since each tank is losing and gaining its liquid at the same rate. Suppose that the amounts of sugar in tank 1 and tank 2 are x_1 and x_2 respectively. The rates of change in the amount of sugar $\dot{x}_1(t)$ and $\dot{x}_2(t)$ are as follows.

$\dot{x}_1(t)$ = Gaining rate - Losing rate

$$= (3\ \ell/min)(0\ g/\ell) + (3\ \ell/min)(\frac{x_2}{10}\ g/\ell)$$

$$-(2\ \ell/min)(\frac{x_1}{20}\ g/\ell) - (4\ \ell/min)(\frac{x_2}{20}\ g/\ell) = -\frac{6}{20}x_1 + \frac{3}{10}x_2$$

$\dot{x}_2(t)$ = Gaining rate - Losing rate

$$= (4\ \ell/min)\left(\frac{x_1}{20}\ g/\ell\right) - (3\ \ell/min)(\frac{x_2}{10}\ g/\ell)$$

$$-(1\ \ell/min)\left(\frac{x_2}{10}\ g/\ell\right) = \frac{4}{20}x_1 - \frac{4}{10}x_2$$

Let's get out of here now.
Are we now going home then?

It's getting dark already.
Let's have a beer over there.

Kirch Hop Pub
Ordy, produce the summary of the amount covered for the three days.
In two large parts, we covered 1st order diff-eq and 2nd order diff-eq.

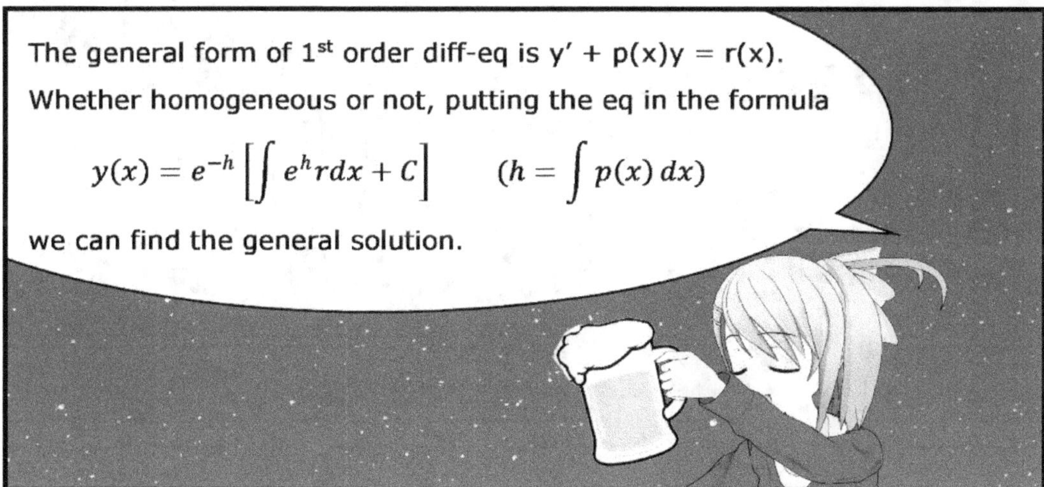

The general form of 1st order diff-eq is y' + p(x)y = r(x).
Whether homogeneous or not, putting the eq in the formula
$$y(x) = e^{-h}\left[\int e^h r\,dx + C\right] \qquad \left(h = \int p(x)\,dx\right)$$
we can find the general solution.

The 1st order diff-eq is applied to many cases regarding velocities.
For instance, it applies to simple Newton Mechanics,
Newton's Law of Cooling, Decay of Radioisotope, Malthusian Law,
Sublimation of a mothball, Dehydration of a comforter, electric circuits, etc.

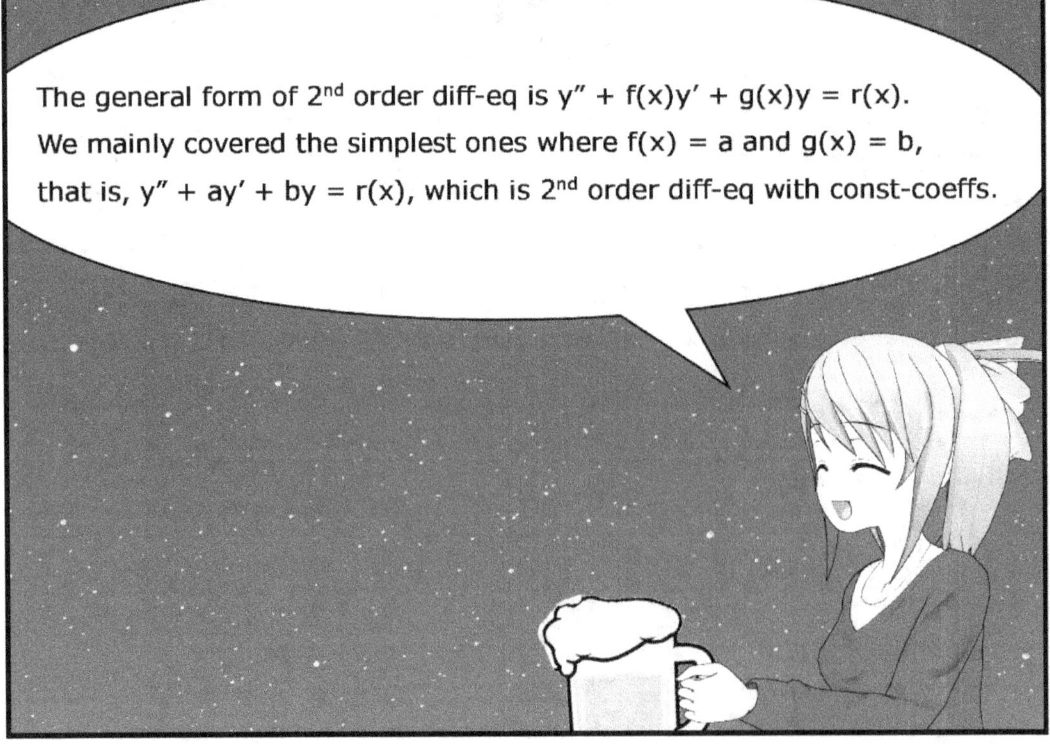
The general form of 2nd order diff-eq is y" + f(x)y' + g(x)y = r(x).
We mainly covered the simplest ones where f(x) = a and g(x) = b,
that is, y" + ay' + by = r(x), which is 2nd order diff-eq with const-coeffs.

When the diff-eq is homogeneous, that is, r(x) = 0, so, y″ + ay′ + by = 0, we assume that the solution y = e^{λx}, find y′, and y″, put them in the diff-eq, get the characteristic eq λ² + aλ + b = 0, and find the roots λ₁ and λ₂. Then, the general solution is yₕ(x) = C₁e^{λ₁x} + C₂e^{λ₂x}.
When the eq is nonhomogeneous, that is, r(x) ≠ 0, the general solution is y(x) = yₕ(x) + yₚ(x), where yₕ(x) is the general solution to the homogeneous diff-eq and yₚ(x) is the particular solution to the nonhomogeneous diff-eq. We find yₚ(x) by means of the method of undetermined coefficients.
To 2nd order diff-eq where coefficients are not constant, there is no general method for the solution, but to some simple ones only we can find the solutions by means of the power series.
We can use Laplace transformations also.

The 2nd order diff-eq gets applied to many cases regarding accelerations.
Typical examples are mechanical systems and electric systems.
In particular, the resonance was the most interesting.

Ordy's skill isn't ordinary anymore.

Besides, there are many others in diff-eq.

Diff-eq
Ordinary Diff-eq
Partial Diff-eq
Linear
Nonlinear
Linear
Nonlinear

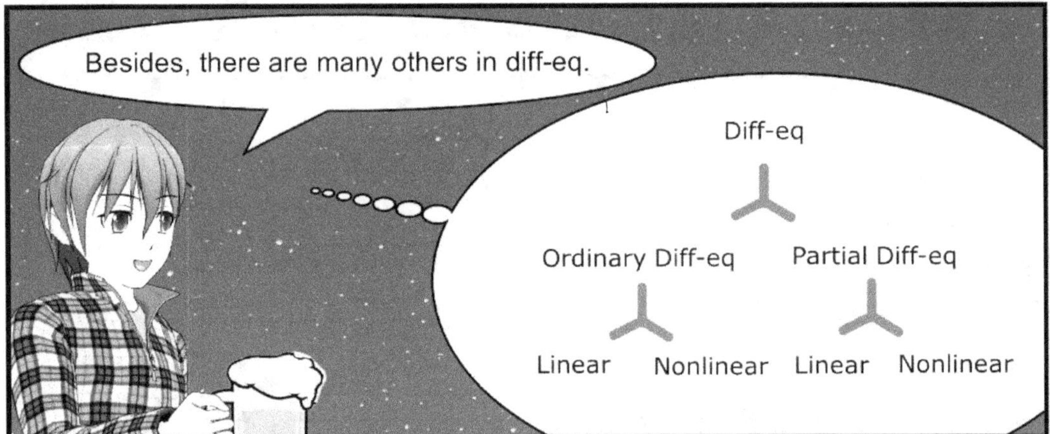

An ordinary diff-eq is a diff-eq for a function
having only one input variable.
For instance, all the diff-eqs covered
in the 3 days are the diff-eqs for y(t)
that is a function of t only.

Also, in the diff-eqs covered over the 3 days, terms in the same kind are added individually in a row. They are called linear diff-eqs.
y″ + ay′ + by = r
We are in the same kind of fish.

Therefore, the diff-eqs covered during the 3 days are linear ordinary diff-eqs.
A nonlinear ordinary diff-eq is an eq where some terms are products of functions as in y″y + ay′y + by = r.

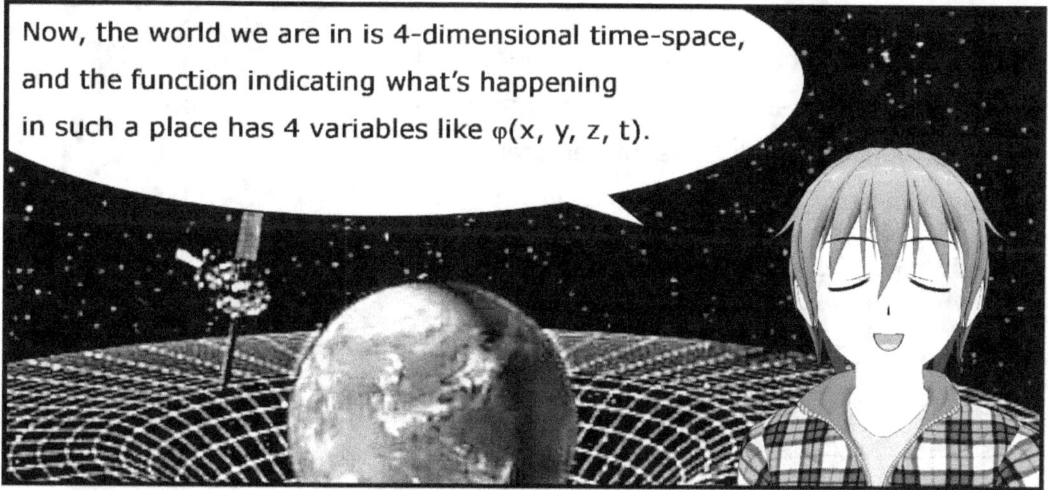

Now, the world we are in is 4-dimensional time-space, and the function indicating what's happening in such a place has 4 variables like φ(x, y, z, t).

Therefore, when we consider a change in the 4-dimensional time-space, we need to make clear what the change is about. For instance, is it:

change with respect to time $\dfrac{\partial \varphi}{\partial t}$?

change with respect to x $\dfrac{\partial \varphi}{\partial x}$?

change with respect to y $\dfrac{\partial \varphi}{\partial y}$?

change with respect to z $\dfrac{\partial \varphi}{\partial z}$?

Those equations that have $\dfrac{\partial \varphi}{\partial t}$, $\dfrac{\partial \varphi}{\partial x}$, $\dfrac{\partial \varphi}{\partial y}$, $\dfrac{\partial \varphi}{\partial z}$, etc.

are called partial diff-eqs.

Because of the recent advances in computer software, we can easily find many application programs available that can produce the solutions immediately if we input diff-eqs only.

Did you have fun for the 3 days, Ordy?
Yes. I hope I have a professor like you.

POP!
See you again!
Oh, my honorable guru! Don't just go away!

What are we going to do with the check for the beer?
Ordy! You said you have an exam today, didn't you!

Oh, dear!
I am getting late
for the exam.

Lynia!
Come with me.

Did you study
for the exam?
Of course!
If it is on
engineering math,
I see no problem!

Outdoor class
works agree with
my constitution,
indeed!